HOW THINGS WORK
FLIGHT

Other publications:

AMERICAN COUNTRY
VOYAGE THROUGH THE UNIVERSE
THE THIRD REICH
THE TIME-LIFE GARDENER'S GUIDE
MYSTERIES OF THE UNKNOWN
TIME FRAME
FIX IT YOURSELF
FITNESS, HEALTH & NUTRITION
SUCCESSFUL PARENTING
HEALTHY HOME COOKING
UNDERSTANDING COMPUTERS
LIBRARY OF NATIONS
THE ENCHANTED WORLD
THE KODAK LIBRARY OF CREATIVE PHOTOGRAPHY
GREAT MEALS IN MINUTES
THE CIVIL WAR
PLANET EARTH
COLLECTOR'S LIBRARY OF THE CIVIL WAR
THE EPIC OF FLIGHT
THE GOOD COOK
WORLD WAR II
HOME REPAIR AND IMPROVEMENT
THE OLD WEST

COVER

A Lockheed TriStar jet touches down at Palmdale, California at dusk. The time exposure, taken by a camera mounted on the airplane's tail fin and triggered by remote control, captures ribbons of runway lights streaming from the jet's nose.

HOW THINGS WORK

FLIGHT

TIME-LIFE BOOKS
ALEXANDRIA, VIRGINIA

Library of Congress Cataloging-in-Publication Data

Flight
p. cm. – (How things work)
Includes index.
ISBN 0-8094-7850-1 (trade)
ISBN (invalid) 0-8094-7851-1 (lib)
1. Flight—Popular works.
I. Time-Life Books. II. Series.
TL546.7.F53 1990
629.132—dc20 90-37365
CIP

How Things Work was produced by

ST. REMY PRESS

PRESIDENT	Pierre Léveillé
PUBLISHER	Kenneth Winchester

Staff for *FLIGHT*

Editor	Elizabeth Cameron
Senior Art Director	Diane Denoncourt
Art Director	Francine Lemieux
Contributing Editors	George Daniels, Peter Pocock, Bryce S. Walker
Assistant Editor	Mitchell Glance
Research Editor	Fiona Gilsenan
Researcher	Hayes Jackson
Picture Editor	Chris Jackson
Designer	Chantal Bilodeau
Illustrators	Maryse Doray, Nicolas Moumouris, Robert Paquet, Maryo Proulx
Index	Christine M. Jacobs
Administrator	Denise Rainville
Accounting Manager	Natalie Watanabe
Production Manager	Michelle Turbide
Systems Coordinator	Jean-Luc Roy

Time-Life Books Inc. is a wholly owned subsidiary of

THE TIME INC. BOOK COMPANY

President and Chief Executive Officer	Kelso F. Sutton
President, Time Inc. Books Direct	Christopher T. Linen

TIME-LIFE BOOKS INC.

EDITOR	George Constable
Director of Design	Louis Klein
Director of Editorial Resources	Phyllis K. Wise
Director of Photography and Research	John Conrad Weiser
PRESIDENT	John M. Fahey Jr.
Senior Vice Presidents	Robert M. DeSena, Paul R. Stewart, Curtis G. Viebranz, Joseph J. Ward
Vice Presidents	Stephen L. Bair, Bonita L. Boezeman, Mary P. Donohoe, Stephen L. Goldstein, Juanita T. James, Andrew P. Kaplan, Trevor Lunn, Susan J. Maruyama, Robert H. Smith
New Product Development	Trevor Lunn, Donia Ann Steele
Supervisor of Quality Control	James King
PUBLISHER	Joseph J. Ward

Editorial Operations

Production	Celia Beattie
Library	Louise D. Forstall
Correspondents	Elisabeth Kraemer-Singh (Bonn); Christina Lieberman (New York); Maria Vincenza Aloisi (Paris); Ann Natanson (Rome).

THE WRITERS

Walter Boyne enlisted in the U.S. Air Force in 1951 and retired as a colonel in 1974 with more than 5,000 flying hours in a score of aircraft. He is the prize-winning author of fifteen books on aviation and automotive subjects, including three novels. Among his non-fiction works are *Boeing B-52—A Documentary History, The Smithsonian Book of Flight* and *The Leading Edge*. His novels include *The Wild Blue* (with Steven L. Thompson), *Trophy for Eagles* and the second book of a trilogy, *Eagles at War.*

Terry Gwynn-Jones served for 32 years as a fighter pilot and flight instructor with the British, Canadian and Australian air forces. In 1975, he teamed with pilot Denys Dalton to set an around-the-world speed record—122 hours and 17 minutes—for piston-engined aircraft. His books include *True Australian Air Stories, On a Wing and a Prayer* and *Farther and Faster.*

Valerie Moolman, a former editor for Time-Life Books, is the author of 45 books, documentary film scripts and dramatic shows on subjects ranging from aviation to social history. She is the author of two books in Time-Life's *Epic of Flight* series and was a contributor to *Understanding Computers* and *Mysteries of the Unknown.*

THE CONSULTANTS

Dr. Tom D. Crouch is Chairman of the Department of Aeronautics at the National Air and Space Museum in Washington, D.C. He holds a Ph.D from Ohio State University and has written several books and articles on the early history of aviation. He is also an avid balloonist.

Dr. Howard S. Wolko is Special Advisor for Technology to the Department of Aeronautics of the National Air and Space Museum. He holds a D.Sc. in Theoretical and Applied Mechanics from George Washington University and has extensive experience in industry, the academics and the Federal Government. He also participated in the development of the X-series of experimental research aircraft.

For information about any Time-Life book, please write:
Reader Information
Time-Life Customer Service
P.O. Box C-32068
Richmond, Virginia
23261-2068

First printing. Printed in U.S.A.
Published simultaneously in Canada.
School and library distribution by Silver Burdett Company, Morristown, New Jersey.

CONTENTS

A World Aloft

The skies are at their most crowded less than 50 feet above the ground. Here countless animals buzz, soar and flap at speeds of up to 75 miles per hour, and ingenious aerodynamic tricks are used to stay aloft. To lift itself off a bowl of potato salad, for example, the domestic housefly must beat its wings furiously—up to 200 times each second. Likewise the American woodcock, a forest-dweller that would rather hide than fly, bounds into the air like a Harrier jump jet. To avoid predators, the "flying" fish manages to launch itself into the air and sail for short distances on its long, wing-like pectoral fins. Humans entered the crowded skies in a modest way on December 17, 1903, when the first successful flying machine, the Wrights' *Flyer*, covered 852 feet in 59 seconds.

Blimp
35-40 miles per hour
Although impractical for commercial air travel, modern helium-filled dirigibles have found a role as flying billboards.

Wrights' *Flyer*
9.8 miles per hour
As his confidence grew with successive flights, Orville Wright pushed the world's first powered airplane to 30 miles per hour.

Woodcock
5 miles per hour
When threatened, the woodcock may reluctantly take to the air, but is the world's slowest flying bird.

Housefly
4 miles per hour
A deft and acrobatic flier, the domestic housefly can even land upside down on the ceiling.

Dandelion seedhead
Borne on the wind
The parachute-like seedhead of the dandelion weed is actually lighter than the air itself. When the relative humidity of the air rises above 70 percent, the seedhead gets heavier and floats to the ground.

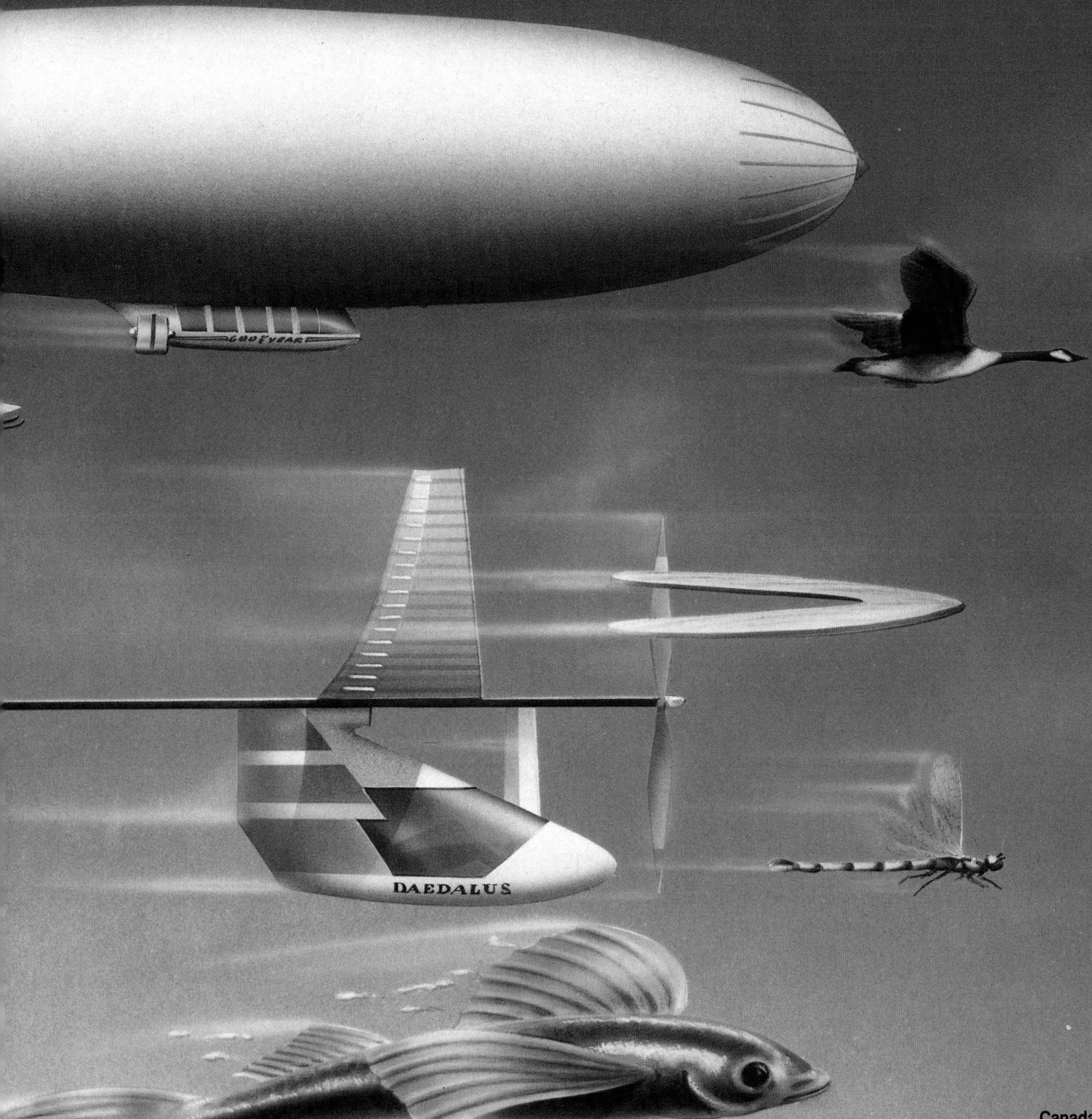

Canada goose
60 miles per hour
Strong chest muscles and long, pointed wings that can span 6 feet from tip to tip enable Canada geese, *Branta canadensis*, to fly hundreds of miles at up to 60 miles per hour.

Boomerang
45-55 miles per hour
In flight, a boomerang spins end over end 10 times per second. Its elliptical flight seldom exceeds 12 seconds.

Daedalus
18 miles per hour
This 70-pound aircraft, piloted by an Olympic cyclist, holds the distance record for human-powered flight: 72 miles.

Dragonfly
60 miles per hour
Engineers studying the dragonfly's use of airflows have clocked the insect at 60 miles per hour. This remarkable flier has inhabited the planet for 250 million years and, for its size, generates three times the lift of the most efficient aircraft.

Flying fish
35 miles per hour
Flying fish skim along the surface of the water to gain momentum, then lift off a wave and glide up to 150 feet on wing-like pectoral fins. They have been clocked aloft for 42 seconds.

Beyond 90 miles per hour, only the peregrine falcon remains from the animal kingdom. At these speeds, for both birds and flying machines, lift is produced by thin, tapered wings called airfoils. Air must travel farther and faster over the curved upper surface of the airfoil, creating two different pressure areas that lift the wing. A helicopter looks different from an airplane, but its whirling rotor blades are in fact spinning airfoils.

Flying disk
74 miles per hour
The familiar spinning plastic disk is a sophisticated aerodynamic toy.

Hockey puck
85-90 miles per hour
When a professional hockey player winds up and takes a slapshot, the vulcanized rubber disk streaks toward the goal at up to 90 miles per hour.

Helicopte
170 miles per hour (cruising
Helicopters usually sacrifice spee for maneuverability and are mor difficult to fly than a plane

Cessna Twin
163 miles per hour (cruising)
This popular business plane has a top speed of 245 miles per hour and a range of more than 1,200 nautical miles.

Voyager
122 miles per hour (cruising)
Voyager entered aviation history in 1986 by flying nonstop around the world on a single tank of fuel.

Boeing 747
580 miles per hour (cruising)
Four turbofan engines and a wing surface area greater than a basketball court are necessary to lift this 390-ton jumbo jet.

Executive jet
509 miles per hour (cruising)
The Gulfstream III carries eight passengers and a crew of three at speeds close to those of commercial jetliners.

F-86E Sabre
690 miles per hour
In the hands of a hot pilot, this Korean War-era fighter could approach the speed of sound in steep dives. A classic fighting jet, the Sabre was powered by a turbojet engine that produced 5,900 pounds of thrust.

Peregrine falcon
217 miles per hour (diving)
Fastest animal in the world, the peregrine falcon folds its wings to dive literally at breakneck speed—the force of impact breaks the neck of its prey.

Golf ball
170 miles per hour
Hit off the tee with a No. 1 wood, a 1.6-ounce golf ball can reach a speed of 170 miles per hour. Its dimpled cover increases the ball's distance and accuracy. Professional golfers can drive the ball more than 300 yards.

Concorde
Mach 2
The world's first commercial supersonic flier, the Concorde can fly 3,050 miles without refueling. Cruising at a maximum altitude of 60,000 feet, it makes the trip between New York and London in three hours and forty-five minutes.

Bell X-1
Mach 1.07
Legendary pilot Chuck Yeager broke the sound barrier aboard "Glamorous Glennis" on October 14, 1947. His top speed was Mach 1.45, or 967 miles per hour.

McDonnell Douglas F-15 Eagle
Mach 2.5 (maximum)
With an airframe able to withstand 9 Gs, the 81,000-pound Eagle is powered by two engines with a combined thrust of 48,000 pounds.

Fighter pilots have known for decades that shock waves begin to form around any object traveling near the speed of sound *(page 122)*. But that speed is not a constant—at sea level it is about 760 miles per hour, while at 40,000 feet it drops to about 660 miles per hour.

To avoid ambiguity, scientists adopted the Mach number, named for the 19th-Century Austrian physicist who first measured the speed of sound. An object's Mach number is equivalent to its speed divided by the speed of sound at the object's altitude.

At this threshold, high temperatures and pressures make a conventional aircraft difficult to handle and put dangerously high stresses on its structural materials. Beyond Mach 1.05, the shock waves fold back over the aircraft and the "ride" smooths out as the plane reaches supersonic speed.

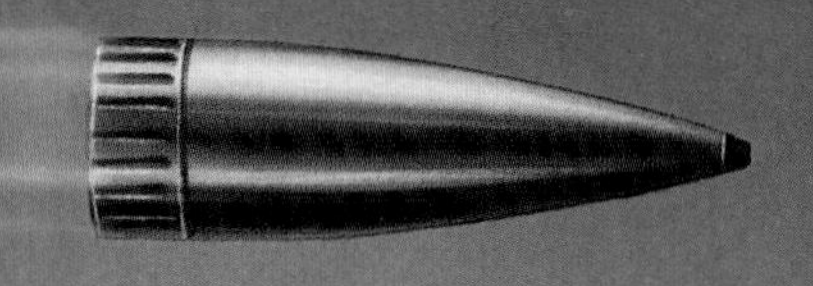

.22 bullet
Mach 2.4
Muzzle velocities vary with ammunition load and muzzle length; the 22-250 Remington rifle bullet reaches speeds of Mach 3.5.

High-speed civil transport plane
Mach 3+ (cruising)
Scheduled for service by the year 2000, the so-called Orient Express will fly from Los Angeles to Sydney, Australia, in 2 1/2 hours and have a nonstop fuel range of 7,000 miles—twice that of the Concorde.

Lockheed SR-71
Mach 3.2 (cruising)
Featuring an airframe of titanium and heat-resistant plastic, this reconnaissance plane is the world's fastest jet.

Space shuttle
Mach 21 (reentry)
The shuttle reenters the Earth's atmosphere at more than 15,000 miles per hour. Protective tiles on the belly of the craft absorb the intense heat created by this friction.

THE WONDER OF FLIGHT

The horizon tilts and the green Earth far below swirls in response to the aircraft's sprightly movements. Rushing air snatches at the pilot's clothes and streams over wings of colorful sailcloth. At a touch of the throttle, the flying machine leaps ahead under the impetus of its spinning propeller. This ultralight aircraft, an ungainly apparatus made from 200 pounds of fabric and aluminum tubing, is one modern expression of the age-old human dream of flying. High above—barely visible as a silver speck creeping across the sky—is another: a huge passenger jet, seven miles up and bound for another continent at 600 miles per hour.

And there are many more. Humans have only recently puzzled out the principles of aviation and created devices that climb into the skies, but those devices have evolved rapidly and taken innumerable forms. Some, such as balloons, are lighter than air, which buoys them up as water supports a floating cork. Others, such as gliders and airplanes, are heavier than air, and stay aloft because of a force, called lift, generated by their movement through the air.

A means of staying in the air is only one of the essential elements of human flight; the development of flying machines has been shaped by other needs as well. The demand for ever-greater speeds has led to powerful, lightweight engines. Maneuvering these faster aircraft, in turn, calls for increasingly complex control systems: Where early pilots shifted their weight to change course, modern fliers activate guidance computers with the slightest of hand movements. Navigation, once a matter of following railroad tracks from town to town, is now possible without a glance at the ground; pilots instead rely on sophisticated electronic equipment and signals from satellites.

Nor is navigation simply wayfinding: The vast expanses where aviators once found only birds and clouds are now so crowded with traffic that their use must be rigorously regulated to prevent collisions.

Cruising about 1,000 feet above the Swiss countryside, the pilot of an ultralight aircraft surveys the drifting passage of a hot-air balloon. A 30-horsepower engine, with a cruising speed of about 40 miles per hour, powers this "lawn chair with wings." Strong, lightweight materials contribute to its safety and economy.

AIR AND ITS MOVEMENT

All this activity takes place in a realm that is itself highly active. The regions where aircraft operate are filled with winds and weather systems, ranging from gentle zephyrs to vast weather fronts, roaring arctic expresses, and formidable jet streams. All result from the turbulent motions of the molecules that make up the Earth's atmosphere. Countless air molecules—about 77 percent nitrogen and 21 percent oxygen with traces of other substances, including water vapor—move about freely, continuously jostling each other and occupying any space open to them. Everything that flies, from bird to balloon to jet plane, relies on interaction with this sea of air to stay aloft.

An understanding of the true nature of the atmosphere, and some of the forces involved in flight, began to emerge in the 17th Century with the work of Evangelista Torricelli, an Italian scientist who studied movement and pressure in fluids—a category that includes gases as well as liquids. In 1643, Torricelli invented the barometer, a device for measuring air pressure. Using simple glass tubes and bowls of mercury, he was able to determine the weight of a narrow column of air reaching from the Earth's surface to the outer edge of the atmosphere—about 14.7 pounds for a column one inch square. Compacted by the weight of overlying air, the atmosphere is most dense at the bottom. The pressure diminishes rapidly with altitude, falling to half its sea-level value about three-and-a-half miles above the surface of the Earth; 90 percent of the atmosphere is below the height of ten miles.

This relatively air-rich layer of the atmosphere, called the troposphere, is the region where clouds form and weather conditions develop. It is also the arena for natural flight, ranging from insects that flutter no more than a few feet from their place of birth, to arctic terns that make an annual round-trip migration of 22,000 miles, and bar-headed geese that soar as high as 30,000 feet as they pass over the Himalayas. Human flight, too, is generally limited to the troposphere, although long-distance jets sometimes venture into the next layer, known as the stratosphere. Manned balloons have explored the upper reaches of the stratosphere, which extends to a height of 30 miles; rocket-powered aircraft have gone beyond it. For the most part, though, flight is confined to altitudes below 80,000 feet, where the atmosphere is thick enough to feed engines and provide lift to the wings.

A SLICE OF THE SKY

As altitude increases, air becomes more rarefied and temperature drops, affecting both natural and mechanical flight. Insects and birds generally fly below 1,000 feet where oxygen is abundant and ambient temperature tolerable. Humans can fly up to 10,000 feet without specialized equipment. Above that height, planes must be equipped with pressurized, temperature-controlled cabins and supercharged engines that can suck in enough of the thin air to power the plane.

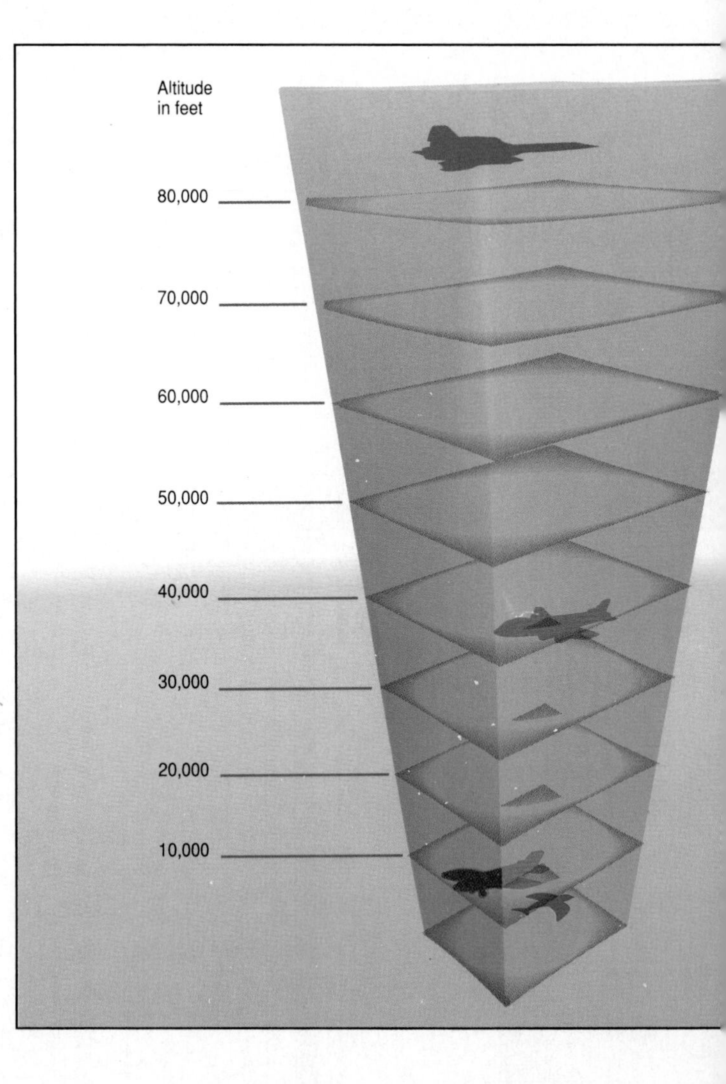

BORNE ON THE WIND

The lift that keeps airplanes aloft is only one of the aerodynamic forces that act on a body moving through the air. Simple air resistance, for example, slows any moving object by imposing a force in the direction opposite to the motion. In the case of a falling body, lift partly counteracts the force of gravity, which causes the object to accelerate—gaining speed at a steady rate. Air resistance increases with the object's speed, so that eventually the upward force generated by the air matches the downward pull of gravity, and the object stops accelerating. It then falls at a constant speed, called terminal velocity. The terminal velocity of a stone is relatively high, since it begets relatively little air resistance compared to its weight; a feather, on the other hand, falls more slowly because its broad surface creates a far greater air resistance in proportion to its negligible weight.

Lift is a more complex phenomenon. It is produced by the motion of a specially shaped surface through the air, and acts in a direction perpendicular to the motion. Thus if a lift-generating body moves horizontally, its lift works to counteract gravity. Without any motive power, a body cannot produce enough lift to overcome gravity and stay aloft indefinitely, but the force does act to slow the descent and allow the body to follow a sloping path to the ground—the graceful action known as gliding.

Nature exploits these forces in myriad ways. Many plants extend their range by distributing seeds that have evolved into aerodynamic shapes. The tiny seeds of the dandelion, suspended from downlike clusters, fall to the ground so slowly that the least breeze can sweep them up and carry them for miles. Other seeds have true wing shapes that generate lift, allowing them to glide long distances. The seed of the tropical palm *Zanonia macrocarpa*, for instance, is so well formed that it inspired the wing shape of an early biplane; it bears a remarkable similarity to the design of an advanced military aircraft, the Northrop B-2 Stealth bomber. The familiar maple seedpod spins as it descends like a tiny helicopter rotor, its wings generating lift that slows its fall and gives the wind time to carry the seed away from its parent tree.

Animals, too, use aerodynamic forces in their quest for survival: flying squirrels, lemurs, snakes, lizards, and—perhaps best-known of all—flying fish. All of these creatures have one thing in common. Though frequently airborne, they do not really fly—at least, not as birds, bats and insects fly, staying aloft indefinitely under their own power. Instead, they use specialized surfaces of their body to glide. The North American flying squirrel, for example, spreads a sail-like membrane between its front and back legs, enabling it to soar from lofty perches for distances of up to a hundred feet at a time. To prepare for landing, the squirrel arches its body, letting the membrane balloon outward to catch air and slow its fall.

A GASEOUS MIX

Gas molecules dance about in this rendering of an iota of air, magnified more than a millionfold. Nitrogen (purple) *and oxygen* (green) *predominate, with water vapor, carbon dioxide and other gases making up the rest. Objects moving through this dense soup of molecules create currents and eddies that produce a variety of aerodynamic forces.*

CANVAS BRAKES TO BREAK A FALL

The nylon canopy that blossoms over the head of a parachutist performs the same function as the arched body of the flying squirrel, creating additional air resistance to cut the speed of descent. Expanded by the rushing air as it opens, the parachute initially develops a braking force that exceeds the pull of gravity, thus causing an actual deceleration. As the velocity dwindles, so does the strength of the air resistance, until it is finally in precise balance with the force of gravity. At this point, with no overall force acting on the parachute and its passenger, the descent continues at about 15 miles per hour without speeding up or slowing down.

The first parachutes, in the late 18th Century, were made like umbrellas. Thin wooden ribs supported vast expanses of canvas, some measuring as much as 195 feet across. The first person to entrust himself to one of these monumental con-

The jump
A skydiver leaps from a small platform outside the open door of a light plane at an altitude of about 10,000 feet and a speed of about 100 miles per hour. Spreading his arms and legs and arching his back to achieve aerodynamic stability, the skydiver begins a free fall.

The fall
Air resistance quickly slows the skydiver's forward movement, and limits the speed of descent to 120 miles per hour. When a wrist altimeter registers an altitude of 2,500 feet, the parachutist pulls the rip cord. A small auxiliary parachute called the pilot chute pops out, fills with air, and pulls the main parachute out of its pack.

The shock
The parachutist experiences a tooth-rattling jolt called opening shock as the main canopy fills with air, abruptly cutting the descent velocity from 120 to about 15 miles per hour. If the main parachute malfunctions, the skydiver quickly deploys a reserve chute; it can be effective if it opens as little as 200 feet above the ground.

This sequence of photographs displays a typical jump of a sport parachutist. Novices do not open the parachute themselves; a static line attached to the aircraft deploys it automatically.

traptions was Louis-Sebastien Lenormand, a French physicist, who drifted safely to the ground from the tower of the Montpellier Observatory, France, in 1783. Two years later another Frenchman, Jean-Pierre Blanchard, used a hot-air balloon as the jumping-off point for the first test of a parachute of his own design. The prudent Blanchard, however, delegated the actual descent to a dog. The wisdom of this decision was borne out in 1793, when Blanchard finally undertook his first parachute jump and broke his leg.

These early, unwieldy parachutes were little more than curiosities, especially in the absence of a way to get into the sky in the first place. As more flying machines like Blanchard's balloon began to appear, their very existence stole some of the mystique of the parachute. Lighter-than-air flight, after all, was more sustainable than a parachute drop, and held the potential for travel as well. The advent of powered flight, however, did create a practical application for the parachute, as a potential lifesaver in the event of midair mechanical failures. The continued development of powered aircraft eventually led to other parachute applications: The chute became a useful instrument for dropping supplies and people—particularly troops—into otherwise inaccessible areas. By the late 20th Century it also evolved into a piece of sporting equipment *(above).* In contrast to its gargantuan forebearers, a modern sport parachute stows neatly into a small pack; even when open, its nylon canopy is only about 25 feet across.

FROM FALLING TO SOARING

The flying machines that would eventually overshadow such novelties as the parachute passed through a long and arduous evolution. Early experimenters had to solve the fundamental problem of obtaining lift, which is only available when air is moving over a lifting surface, such as a wing. One way to achieve this is to equip the craft with a power source that will move it through the air. But as long as engines remained too heavy—as they did until the Wright Brothers' historic flight at the beginning of the 20th Century—powered flight remained an attribute unique to the animal kingdom. Two other lifting techniques were at hand, though. Would-be fliers could tap the power of gravity by climbing to a high place, like a mountain or hillside, then jumping off and using the speed of the fall to get the

The ride
An apex vent in the top of the parachute allows compressed air to escape and permits a steady descent. To achieve the control over direction and speed of descent needed for an accurate landing, the skydiver uses steering lines that spill air out of slots or gores in the chute.

The touchdown
Watching the approaching ground intently, the parachutist bends his knees just before landing. Even after a 10,000-foot descent, the impact is about the same as jumping off a wall eight feet high. At touchdown, the skydiver rolls with the momentum, absorbing most of the shock with his leg muscles, and immediately starts to gather in his chute.

air moving over the wing. Alternatively, they could seek out updrafts—masses of upward-moving air—and other wind patterns that would provide lift to keep the craft in the air.

Birds use both of these techniques to stay aloft with a minimum expenditure of energy. The Andean condor, for example, follows a carefully selected flight path that takes it into winds rushing up steep mountain slopes. After rising effortlessly with the moving air, the condor can then move away from the mountain, gliding at a shallow downward angle and trading altitude for lift-generating speed. Eventually the condor returns to the region of the updraft to take another ride to the top. Other birds, from albatrosses to vultures, use similar tactics in their continuous aerial hunt for food *(right)*. Migratory fowl, too, take advantage of weather patterns; often their flight paths allow them to soar on winds deflected upward by ridges. Some scientists suggest that the characteristic "V" formation of geese is also an energy-saving stratagem that allows every bird but the leader to rest its wing tip on the rising vortex of air displaced by the wing of the bird in front. The hard-working goose at the apex of the "V" periodically falls back into the formation, to be replaced by another, so that the free rides are shared equitably.

Lessons learned from observations of soaring birds found their way long ago into the construction of kites, the first man-made objects that exploited the forces in moving air. Appearing in China more than 2,000 years ago, the tethered fliers were originally used by adults in religious ceremonies and in war, as well as for entertainment. Chinese generals hoisted soldiers aloft on giant kites to observe enemy troops, and the first concrete evidence of kites in Europe is a book written in the 14th Century depicting three soldiers using a kite to drop a bomb on a castle. Only later did kites assume their modern identity as children's toys, although they were still employed at tasks as diverse as fishing in hard-to-reach waters and experiments with electricity.

As varied in use and size as they have been since antiquity, all kites share a reliance upon the simplest aerodynamic principles. The force that holds a kite aloft comes from the resistance of its surface area to the wind. Held at an angle by its string, the kite deflects the moving air downward; in a breeze of constant strength, the forces applied by the wind and the string are in perfect balance, and the kite hangs motionless in the air.

By the late 19th Century these forces were the subject of rigorous experimentation in Europe, where several inventors hoped to bring new generations of kites to war. One leading proponent of man-carrying kites was B. F. S. Baden-Powell, a British Army officer and brother of the founder of the Boy Scouts. His intention was to provide the army with a means of aerial observation. Baden-Powell's first success was in 1894, with a giant kite made of cotton cambric stretched over a bamboo frame. Thirty-six feet high, the kite had enough power to lift a man, although safety considerations limited flights to no more than ten feet above the ground (Baden-Powell usually took along a parachute on his own test flights). The next year Baden-Powell patented a new design, which used several kites about ten feet square, linked one above the other by ropes. The apparatus, which Baden-Powell called a Levitor, used four to seven small kites at a time, depending on the wind conditions. In moderate to strong winds, the Levitor proved capable of lifting the inventor as high as 100 feet.

SOARING AT SEA

The swooping flight of long-winged seabirds takes advantage of slower wind speeds near the surface of the open sea. With the wind at its back, an albatross descends from a height of about 50 feet, gaining speed and distance. Mere inches from the wave tops, it wheels into the weaker surface breeze, then uses its momentum to climb back through stronger winds before turning for another downwind dive. Using this technique, the albatross can patrol the ocean for hours without flapping its wings.

COASTING ON CURRENTS

The broad wings of eagles and condors allow the heavy birds to ride aloft on currents of air sweeping upward over cliffs and hillsides. The gliding birds tack across the updraft, steadily gaining altitude as they go.

RIDING THERMALS

Land birds with broad, rectangular wings frequently use thermals—columns of hot, rising air—to gain altitude. Entering the thermal near the ground, a vulture circles within the column, gaining several thousand feet with very little effort.

Big kites soon found other functions. In 1901, Italian inventor Guglielmo Marconi used a Levitor to raise the receiving antenna for his first transatlantic wireless test at St. John's, Newfoundland. A strong wind tore apart one kite, but a second fared better, carrying the antenna to a height of 400 feet, and Marconi received the message he was listening for. For a brief period in the 1900s, the British Army adopted a system of observation kites (using a design that was safer and more stable than Baden-Powell's); during World War I, the German navy sent similar man-lifting devices to sea with their submarines, vastly increasing the U-boats' chances of discovering distant prey. Like parachutes, however, kites were suited only for a few special jobs. And while kite-builders struggled to perfect their tethered creations, other inventors were pouring their energy into machines that would not merely resist the wind, but ride upon it.

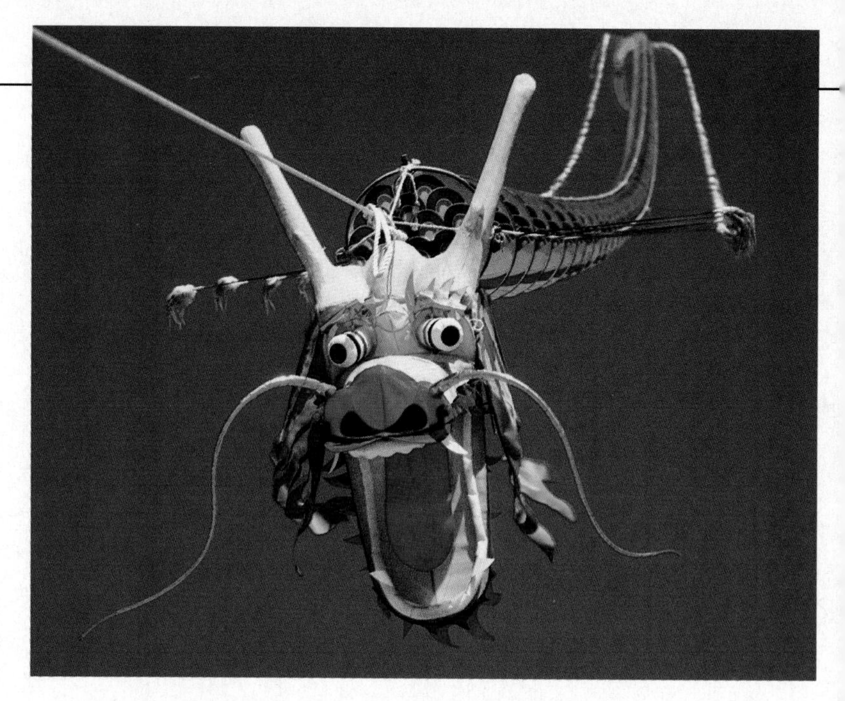

The fearsome face of a dragon supports a long, undulating tail on a classic Chinese kite. Made of hand-painted paper covering a split-bamboo frame, it is little different from kites that flew 2,000 years ago.

BREAKING THE TETHER

Otto Lilienthal was one of this new breed of experimenters. A distinguished German engineer, Lilienthal was captivated by the sweeping flight of seagulls he encountered while installing a foghorn of his invention in German lighthouses. He studied birds in meticulous detail, eventually publishing an authoritative book drawing connections between natural and artificial flight. His aviation research was equally painstaking, beginning with a series of kite experiments in the 1870s and progressing to free-flying machines with wings modeled after those of soaring birds. In the course of five years in the 1890s, Lilienthal flew 18 different kinds of gliders, taking careful notes on their aerodynamic qualities.

He launched his flights from the top of a 50-foot hill, built specially for the purpose in an open area near Berlin. The hill was conical, which allowed Lilienthal to fly directly into the wind no matter which direction it blew, thus increasing the lift-generating movement of air over the curved, fabric-covered wings of his fixed-wing gliders. Half-sitting on a trapeze so that his head and shoulders were above the wings, Lilienthal would step off the hill to begin a long, gentle glide to the ground. Maneuvering the craft by shifting his weight, the inventor regularly flew for distances up to 750 feet, becoming the first man to achieve sustained, controlled flight in a heavier-than-air machine.

Lilienthal paid dearly for his achievement. He was piloting one of his most reliable gliders on a summer's day in 1896 when a sudden gust of wind brought him to a standstill in midair. One wing lost its lifting power, dropped sharply, and the glider sideslipped to the ground. Lilienthal, his spine broken in the crash, died the next day. His work outlived him, however. The body of aerodynamic data generated by his experiments was an important step in the development of powered flight, which became a reality less than a decade after his death. And even after aircraft with engines began winging their way through the skies, the memory of

Eighty feet from tip to tip, the slender wings of a sailplane carry it on air currents at play above the Rocky Mountains.

Blaník

Lilienthal's graceful gliders lived on in a quieter, more contemplative style of flying, embodied in the modern sailplane.

This sporting craft, with its long, tapered wings and streamlined, glassy-smooth fuselage, is a paragon of specialized efficiency in design, structure and materials. Built to produce maximum lift, it can soar for hours when conditions are right, covering hundreds of miles at speeds up to 150 miles per hour in a dive. Every flight, however, is a battle of human wits against nature; to stay aloft, a pilot must find an updraft and use it as an elevator to gain altitude. The pilot looks for visual cues—birds spiraling, leaves blowing upward, rising clouds—and then steers for them. But updrafts are not always available, and they sometimes disappear; the prudent pilot watches the rate-of-climb indicator to monitor lost altitude, while keeping an eye on the home field or an alternate landing spot within gliding range.

LIGHTER THAN AIR

The rising masses of warm air that prove useful to soaring birds and glider pilots are instrumental to another, completely different kind of flight. Captured in a leakproof bag, warm air retains its tendency to rise, and can be harnessed to lift weight as long as it is warmer, and thus lighter, than the air around it.

This simple principle occurred to French paper-maker Joseph Montgolfier one evening in 1783 as he watched the fire in his hearth. Struck by the potential lifting power of heated air, Montgolfier made a bag of fine silk and lit a fire under it. The bag filled up with hot air and rose to the ceiling. In June of that year, aided by his brother, Montgolfier demonstrated the effect in public by raising a huge cloth bag filled with hot, smoky air rising from burning wood and straw. When this primitive balloon actually proved workable, the brothers attached a basket and selected the world's first balloonists—a duck, a sheep and a cockerel. The animals landed unharmed after an eight-minute voyage.

The following month—November 1783—two human volunteers ascended in a Montgolfier balloon equipped with a brazier. The fledgling fliers took along a bucket of water and sponges in case the fire got out of control, but the first manned balloon ride was a complete triumph: The balloon rose 500 feet and floated above the rooftops of Paris for 25 minutes.

Even as the Montgolfier brothers were developing their hot-air balloon, another experimenter was working on a completely different approach to lighter-than-air flight. Professor Jacques A. C. Charles filled a small bag of rubberized silk with hydrogen, a gas only seven percent as heavy as air. Launched in Paris, it flew for about 45 minutes before coming down in the village

ARCHIMEDES' PRINCIPLE

The ascent of a hot-air balloon illustrates Archimedes' Principle, which states that any object suspended in a fluid experiences a buoyant force equal to the weight of the fluid it displaces. A typical balloon is subjected to an upward force of about 8,000 pounds—the weight of the air it displaces. When the air inside the envelope is heated, it expands and about one-quarter escapes through the bottom opening, reducing the overall weight of the balloon. Now the buoyant force exceeds the weight, and the balloon rises.

of Gonesse, about 15 miles away. The balloon met an unfortunate end: It was torn to shreds by pitchfork-wielding villagers, intent on destroying what they believed was a device of the devil. Undeterred, Charles built a larger, man-carrying balloon, which he flew just a month after Montgolfier's flight. The enthusiasm for ballooning had grown so great that Charles's takeoff from the Tuileries Gardens was witnessed by an estimated 400,000 people. The 28-foot-wide balloon, remarkably similar to modern gas-filled designs in terms of construction, valves and ballast, was a complete success, carrying Charles and a friend 27 miles in a two-hour flight.

STEERING IN THE SKIES

While balloons quickly proved their ability to carry people into the sky and over distances, they suffered the drawback of almost complete unpredictability. The only possible destination for any given balloon trip was somewhere directly downwind from the launch site. Hence inventors immediately turned to the problem of creating a balloon that could be steered under its own power: a dirigible. They were thwarted, however, by the materials and equipment available. The first airships were essentially elongated balloons, filled with hydrogen, trussed into shape like corseted whales, and equipped with primitive engines and rudimentary steering devices. These early airships were particularly prone to losing their aerodynamic shape, sagging limply with any decrease in the pressure of the gas that kept them inflated.

An inkling of a solution to this problem appeared in 1900, with the flight of the first rigid airship. Designed by Count Ferdinand von Zeppelin, a retired German cavalry officer, the dirigible boasted a rigid aluminum frame and a streamlined fabric envelope surrounding multiple gas-filled lifting compartments inflated with hydrogen. At 420 feet long and almost 40 feet in diameter, it was far larger than previous airships, and held enough hydrogen to lift 27,000 pounds. Structure, engines and ballast weighed so much, however, that the payload for the giant's first flight was just 660 pounds.

Count von Zeppelin spent the next decade developing his idea, overcoming business setbacks and a series of disastrous wrecks and fires. Eventually his ships proved powerful and reliable enough for a passenger airline serving a number of German cities to be inaugurated. During World War I, Zeppelins in the service of the German military even carried out bombing raids over England. But dirigible technology reached its peak of development after the war (and after the death of von Zeppelin in 1917), when a new generation of the giant dirigibles established high standards for comfort and elegance in travel. As fast as contemporary airplanes, these rigid airships had far greater range and carrying capacity than heavier-than-air machines. In the late 1920s, the *Graf Zeppelin*, most successful of the German fleet, routinely carried up to 20 passengers across the Atlantic in about 66 hours, pampering the travelers en route with superb cuisine and unmatched vistas of ocean and sky.

The glory days of the giant airships were numbered, however. By the early 1930s many of the dirigibles' performance advantages were surpassed by aircraft. Furthermore, structural deficiencies still lurked in the design of the enormous craft. Their aluminum frames were so large (the *Hindenburg*, star of the fleet in the 1930s, was 804 feet long and 134 feet in diameter) and so lightly built that the ships were extremely vulnerable to bad weather; turbulence could easily stress and distort the fragile structures.

An even greater hazard, however, was the flammable hydrogen used as a lifting gas in most dirigibles. In practical terms, the age of the rigid airship ended on May 6, 1937, at Lakehurst, New Jersey, when, for reasons unknown, the hydrogen aboard the *Hindenburg* suddenly burst into flame. The popular ship was just coming in for a landing, and the disaster, which claimed 35 lives, was reported to a stunned world in a live radio broadcast. Newsreel film showed the pride of the skies reduced to a bent, blackened skeleton.

The grim demise of the *Hindenburg* closed a chapter in the history of air travel, although airships continued to hold their specialized place in the lineup of flying machines. Sophisticated gas-control systems have made nonrigid dirigibles more practical; nonflammable helium has made them far safer. Today, airships are routinely employed to patrol vast expanses of ocean, monitor changing weather conditions, haul cargo in and out of otherwise inaccessible areas and, in their most public role, to carry advertising messages as flying billboards.

But airships, like balloons, kites and gliders, have proved to be peripheral to the main business of human flight. Each in its way helped give people a taste of flight, but none can deliver the combination of speed, power and mass access offered by aircraft with engines. With these, humans would conquer the skies.

Preparing a hot-air balloon for flight, a helper holds the envelope open for the flame that heats the air trapped inside. The expanding envelope is already beginning to rise from the ground.

Top deflation port
To collapse the envelope after landing, the pilot pulls a cord that peels back the fabric port; hot air rushes out of the envelope, deflating it almost instantaneously.

Side vent
To make adjustments in altitude or initiate a descent, the pilot uses another cord to open the side vent and spill some of the hot air out of the envelope. When the cord is released, the vent automatically closes.

Envelope
Made of polyurethane-coated rip-stop nylon, the 24 brightly colored panels are seamed together with double rows of stitching reinforced by heavy-duty web tapes. A typical envelope requires 1,000 yards of fabric, more than three miles of thread, and almost half a mile of web tape.

Skirt
The heat-resistant fabric of the skirt funnels hot air into the balloon and protects the burner flames from strong winds.

Burners
Powerful enough to heat 120 houses, each burner is activated when the pilot reaches up and pulls a trigger on the burner blast valve.

Control console
The instrument panel includes a rate-of-climb indicator, an altimeter, and a pyrometer, which registers the air temperature near the top of the envelope.

Fuel tanks
Compressed propane gas is stored in aluminum or stainless-steel tanks strapped inside the basket.

Basket
Traditional wicker may be replaced by aluminum and fiberglass.

RIDING CURRENTS OF AIR

Nearly 60 feet in diameter, a typical hot-air balloon displaces more than 100,000 cubic feet of air. Drifting with the wind at an altitude of 400 feet, the colorful orb can stay aloft for hours at a time.

Vessels of the Air

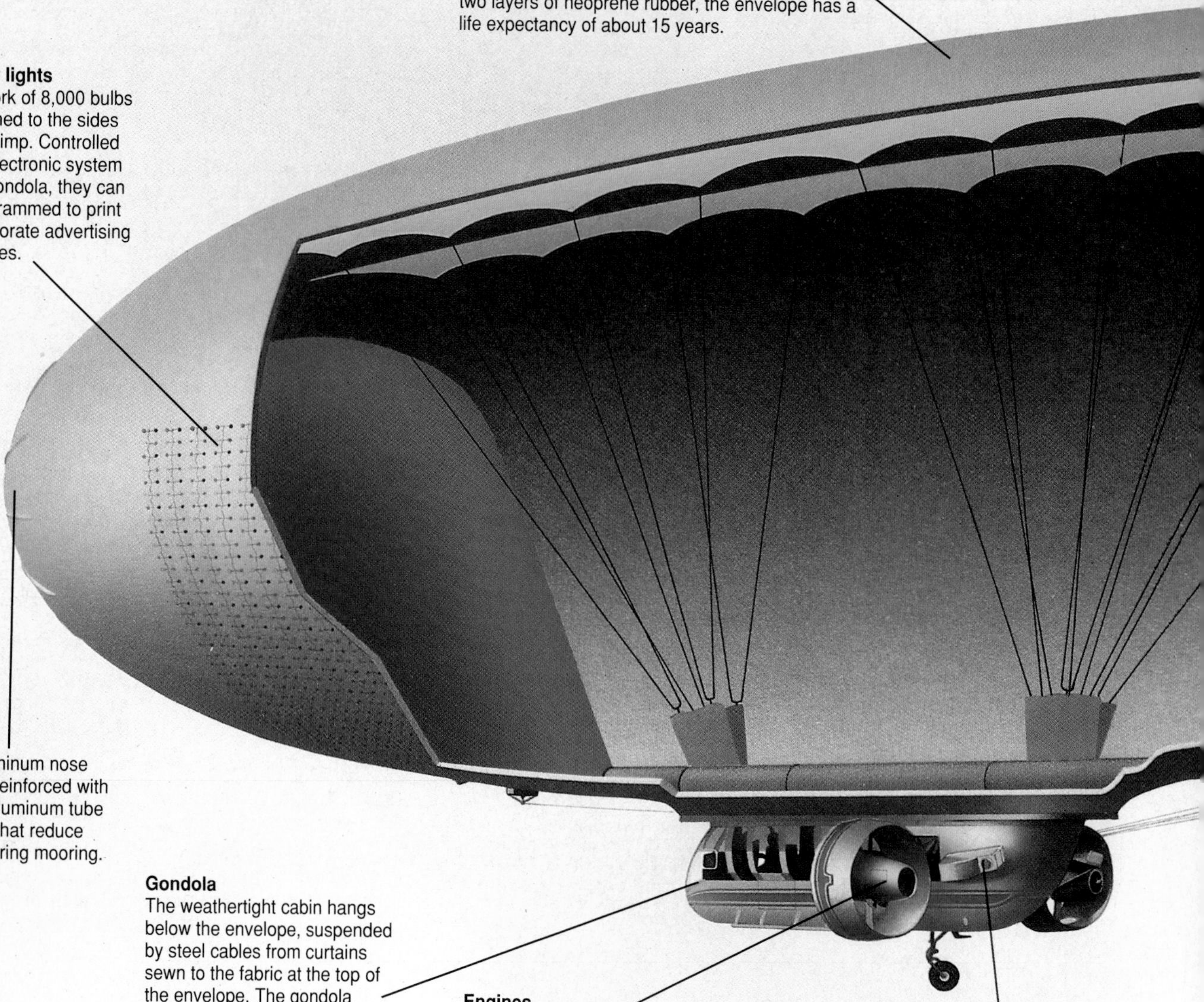

Envelope
The flexible outer casing is about 200 feet long, 50 feet wide and 60 feet high. Made from approximately 2,800 square yards of synthetic fabric coated with two layers of neoprene rubber, the envelope has a life expectancy of about 15 years.

Display lights
A network of 8,000 bulbs is attached to the sides of the blimp. Controlled by an electronic system in the gondola, they can be programmed to print out elaborate advertising messages.

Nose
The aluminum nose cone is reinforced with hollow aluminum tube battens that reduce strain during mooring.

Gondola
The weathertight cabin hangs below the envelope, suspended by steel cables from curtains sewn to the fabric at the top of the envelope. The gondola houses the pilot's controls and navigational instruments, and accommodates a pilot and up to nine passengers.

Engines
Two 420-horsepower turboprop engines propel the airship. The pilot can reverse the thrust of the engines to stop in midair or to back up.

Air scoops
Hanging behind each of the turboprop engines, the air scoops are opened to force air into the ballonets.

A blimp is a steerable, powered balloon that uses internal gas pressure to maintain its shape; the buoyancy of the gas keeps it aloft. A blimp like the one shown here (modeled after the Goodyear blimps) weighs about 12,000 pounds, not including pilot, passengers and cargo, and holds about 250,000 cubic feet of helium—a nonflammable gas seven times lighter than air.

At the takeoff command of "Up ship," the two engines roar to full power. The airship noses up at a startling angle, beginning a laborious ascent to its cruising altitude, about 1,000 feet. Even after the airship levels off, the wind causes it to pitch gently up and down, like a ship riding ocean waves. Throttled back, the engines are very efficient: An airship can fly eight hours a day for almost a week on the amount of fuel it takes to taxi a jumbo jet from ramp to departure gate.

To land, the pilot points the nose down and brings the engines back to full power, driving the airship toward the ground at 30 miles per hour. The course is squarely into the wind, to prevent the nose from being blown sideways as the blimp loses speed. The ship slows almost to a stop, ground crew quickly load on ballast, then a spindle on the airship's nose engages with latches on a tall mooring mast.

Anchored only at this point, the airship is free to rotate 360 degrees around the mast as the wind changes direction.

Fins
Like the feathers on the tail of an arrow, four fabric-covered fins stabilize the blimp in flight. Anchored to the envelope, the fins are supported by guy wires.

Ruddervators
Movable panels on the trailing edges of the fins, controlled by wires from the gondola, serve to steer the blimp in flight.

Ballonets
Two rubber air chambers, at the front and back of the envelope, are inflated or deflated as necessary to maintain normal pressure inside the ship. Each ballonet holds up to 30,000 cubic feet of air, but they are normally only 10 to 30 percent inflated. The ballonets can also be used to balance the airship.

Ballonet valves
Each ballonet has two valves that automatically open and close to maintain a pressure of less than one pound per square inch inside the envelope.

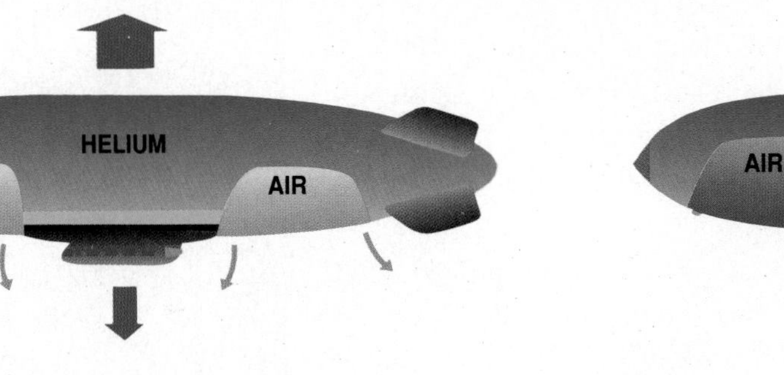

ASCENT
As the blimp climbs, the helium expands and the ballonet valves automatically open to vent air and relieve the excess pressure without releasing expensive helium. The ballonets can compensate for the expanding helium up to an altitude of about 10,000 feet. At that level, the ballonets are completely collapsed; to climb higher, the pilot must release helium.

DESCENT
The helium contracts as the blimp moves to a lower altitude, so the pilot opens the air scoops and the twin turboprop engines force air into the ballonets. The expanding ballonets keep the pressure constant, maintaining the shape of the envelope. Auxiliary blowers in the air scoops supplement the engine-driven air when a rapid descent requires faster inflation.

SHAPE AND POWER

It is a moment every U. S. Navy carrier pilot must learn to face—the blurred, gut-wrenching instant of takeoff, when his F-14 Tomcat leaves the deck as a 30-ton unguided ballistic missile, hurled skyward by the world's most powerful slingshot.

The flight deck of an aircraft carrier is a maelstrom of organized confusion and violent sound. Planes land and lift off at split-second intervals, and the blast of their jets is so shrill that deck crews must wear ear protectors. Two steam-powered launch catapults shudder and hiss. The grinding noise of arresting-gear engines that power the landing-gear mechanism reverberates through the steel deck. An officer orchestrates traffic toward the catapults; he must launch a squadron of 20 fighter planes, at the rate of one every 45 seconds. From a control tower 140 feet above the rolling sea, an air boss monitors the entire operation, his finger poised to hit the abort button at the first hint of trouble.

The first F-14 is hitched to the forward catapult. A catapult officer checks that its wing flaps are set and signals the pilot to apply full power. The pilot revs up, makes a final check of his engine gauges, braces against his seat and throws a smart salute to the catapult officer.

The catapult is preset to fling the F-14 from standstill to 200 miles per hour, virtually instantaneously. So sudden is the acceleration that it confuses the senses. Pilots swear that their aircraft slow down upon leaving the deck, a misconception that arises because the brain takes several seconds to recover from the "catstroke." Even flight instruments lag behind. Battling gravity, the F-14 dips momentarily toward the waves. Then its 40,000 pounds of jet thrust bite hard, and the critical transition occurs: from uncontrolled missile to flyable aircraft. The pilot, now in full command, points the fighter's needle nose upward. The Tomcat rockets into the sky at a 60-degree angle.

With both throttles jammed full forward, the F-14's afterburners kick in, punching it through the sound barrier. Climbing almost vertically, faster than a .45 caliber bullet, the Tomcat reaches 50,000 feet in two minutes. Leveling off, it passes

A carrier-based F-14 Tomcat revs its engines to full screaming power just seconds before takeoff.

Mach 2—twice the speed of sound—accelerating toward its maximum velocity of 1,650 miles per hour.

What enables this amazing fighter to fly? Few people can gaze at the streamlined beauty of a modern jet as it roars overhead without asking the same question. The answer—or at least part of the answer—can be traced back to the Wright Brothers who, nearly a century ago, took a close, discerning look at the world of nature, and so discovered the secret to powered flight.

THE FLIGHT OF BIRDS

Ever since humans first looked skyward, they have envied the flight of birds. Even in the era of transatlantic flights and space shuttles, aviation scientists still study them. Birds remain the lords of the air—the world's purest and most aerodynamically efficient fliers.

These remarkable creatures are capable of reaching speeds in excess of 100 miles per hour, of making migratory flights halfway around the globe, of soaring effortlessly on the wind for hours, and of performing aerobatics and dogfighting maneuvers that would cause a fighter pilot's head to spin. They serve as living blueprints of the theory of flight. Even the leading-edge flaps on contemporary jets find their counterpart in birds' wings. Indeed, without their example, humankind might still be searching for ways to conquer the air.

Like airplanes, birds were not always efficient fliers. Nature has taken millions of years to adapt them into the flying machines they are today. The discovery of a 150-million-year-old fossil in a Bavarian limestone quarry in 1861 gave science its first known bird. Christened *Archaeopteryx*—old wing—it was as much a feathered reptile as a bona fide fowl, and it provided convincing evidence for Charles Darwin's theory of evolution.

Within the next 150 million years, impelled by the needs of feeding, breeding and survival, birds developed into their modern form. Scientists believe that their

A multiflash photograph of an owl in flight reveals the powerful downstroke of the wings: The tips of the primary feathers are bent upward, flaring out almost at right angles to the wing. As the wing descends, these feathers bite into the air and pull the bird forward. At the end of each downstroke, the wings are quickly brought back and tucked slightly in an oar-like movement that provides additional thrust. On the upstroke, the primary feathers separate to reduce air resistance. Then the whole cycle begins again.

light, hollow wing feathers evolved from thick reptilian scales and their stiff tail feathers, which act as in-flight stabilizers from the bony whiplash tail of their reptilian ancestors. To reduce weight further, sturdy bones suitable for land dwelling evolved into hollow, buoyant structures with paper-thin walls, some braced with cross-struts like the wings of airplanes.

The earliest experimenters with flight tried to mimic the flight of birds by strapping on homemade wings, but a human's bulky, ponderous physique is simply not designed for the task of self-propelled flight. Nor could the human heart cope with the aerobic demand of wing flapping—which, in the humble sparrow, is an astonishing 800 heartbeats per minute.

The pioneer birdmen were victims of a common misconception. For centuries, it was believed that birds "swam" across the sky, propelled by a backward and downward wing stroke. Modern high-speed photography shows quite a different process. A bird's forward thrust comes from the primary feathers at the wing tips, which act almost like propellers. As each downstroke begins, the tips of these primaries twist up at an angle to the rest of the wing. The descending tips thus bite into the air in much the same way as the angled blades of a propeller, pulling the bird forward. At the same time, the rest of the wing provides the lifting surface that keeps the bird airborne.

THE WRIGHTS AND THE TURKEY VULTURE

It was only when early aviators began to focus on gliding rather than flapping that they learned how the curved airfoil shape of a bird's wing generates aerodynamic lift. The first person to light upon this phenomenon was the British visionary Sir George Cayley, often called "The Father of the Airplane." In 1810 Cayley issued a pioneering report, *On Aerial Navigation*, which revealed his studies of crows and other soaring birds. In effect, it marked the birth of the science of aerodynamics. Cayley soon had more to say—on the shapes of airfoils, on movable control

Alula
A small tuft of feathers, originating from the thumb, the alula sweeps forward and separates when a bird lands.

Hand section
Similar to a human hand but with longer and narrower bones in relation to the rest of the arm, the bird's "hand" has just two fused fingers and a thumb that can open and close.

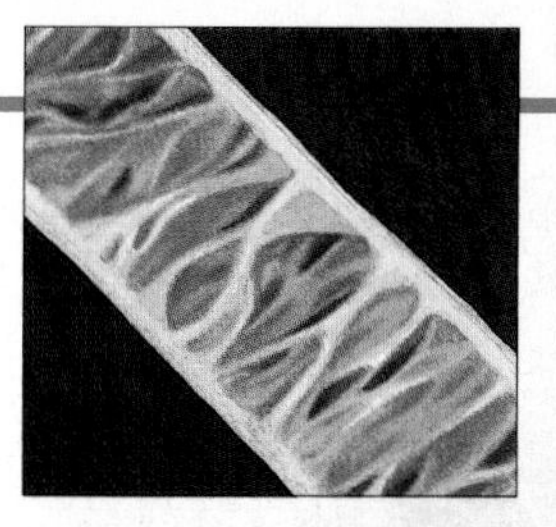

Bones
Filled with air, the honeycombed long bones of flying birds are reinforced by a crisscross of internal struts, much like the frame of an airplane or the stem of many plants.

Primary flight feathers
Stemming from the finger bones of the hand, they are largely responsible for the power for flight.

Wings
Composed of an upper arm, a forearm and a hand section, the wing is attached to breast muscles through a triple-jointed support that lends flexibility and allows the shoulder to be twisted into different positions during flight.

Secondary flight feathers
These flight feathers stem from the outside bone of the forearm and supplement the primary flight feathers.

Tail feathers
Help the wing balance, steer and brake the bird's body in flight.

Breastbone
The breastbone has a deep keel jutting out of its center, anchoring the bird's huge flight muscles.

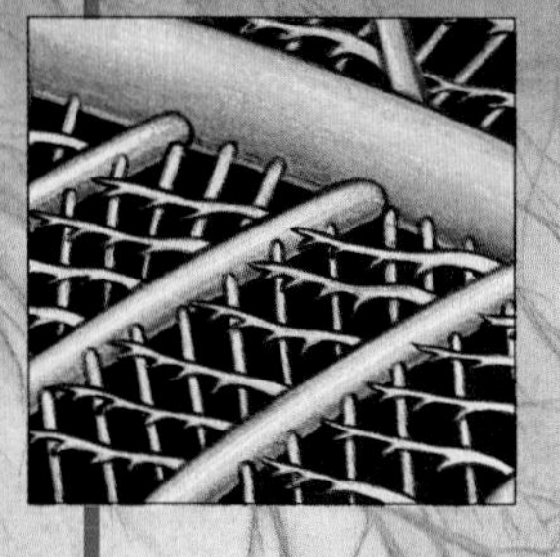

Feathers
More than a million very fine projections, branching off a central shaft, are meshed together by miniscule hooklets, too small to be seen by the naked eye. If the hooklets disengage, the bird simply draws the feather through its beak several times and the projections automatically rehook, just like a zipper.

Built for Flight

Behold the bird—nature's perfect flying machine and the envy of all aircraft designers. Their wings are driven by muscles that are amongst the most powerful in the animal world.

Muscle-powered flight requires a high metabolic rate and a very efficient respiratory system. A bird's unusually large heart pounds at a fantastic rate—up to 1,000 beats per minute in a hummingbird. Its lungs are supplemented by a series of air sacs, spread throughout the body, that supply a continuous flow of oxygen to the powerful flight muscles, and help keep the bird's body temperature at a reasonable level.

To keep weight to a minimum, evolution has rid the bird of excess baggage. For instance, lightweight beaks have replaced heavy jaws laded with teeth. The bones are literally filled with air, crisscrossed by truss-like supports that provide both strength and resilience. The skeleton of a three-pound frigate bird, despite its seven-foot wingspan, weighs just four ounces. This efficiency of form is also apparent in the bird's most important structure. Feathers are one of the lightest yet strongest materials formed by any creature. They serve a dual purpose: providing the surfaces on the wings and tails for lift and propulsion, and insulating the bird against loss of heat and protecting it from the weather.

Although man has perfected designs and developed powerful engines that enable him to travel far faster than any bird, in aircraft that incorporate the latest in digital technology, he has still to attain the refined simplicity and efficiency of two flapping wings.

ANATOMY OF A POWERFUL FLIER

The ubiquitous Columba livia, *commonly known as the rock dove or pigeon, is about 13 inches long and weighs between 10 1/2 to 15 ounces. A strong flier, it is a very familiar sight to city dwellers worldwide.*

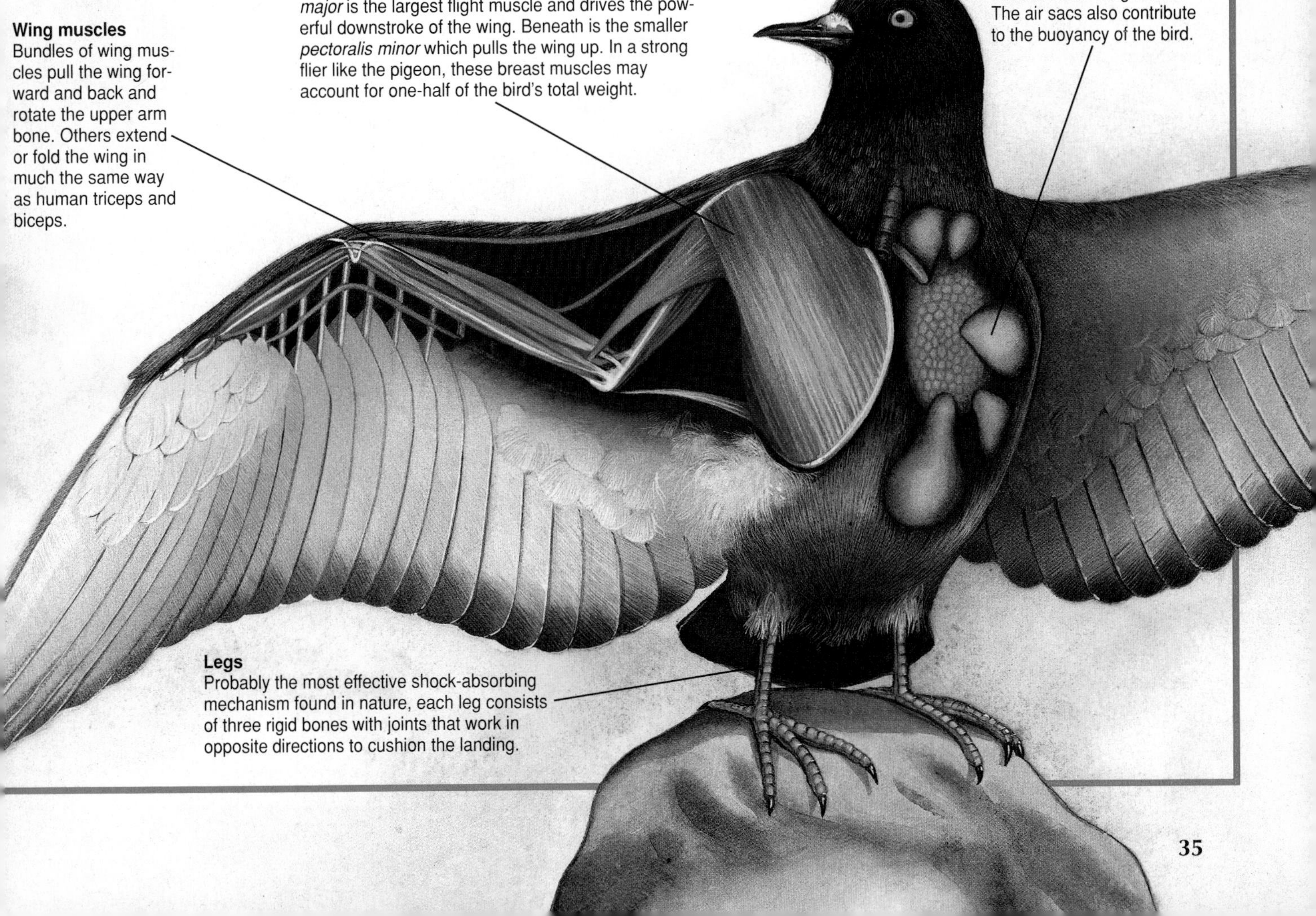

Wing muscles
Bundles of wing muscles pull the wing forward and back and rotate the upper arm bone. Others extend or fold the wing in much the same way as human triceps and biceps.

Breast muscles
Anchored between the breastbone, the *pectoralis major* is the largest flight muscle and drives the powerful downstroke of the wing. Beneath is the smaller *pectoralis minor* which pulls the wing up. In a strong flier like the pigeon, these breast muscles may account for one-half of the bird's total weight.

Air sacs
Connected to the lungs and interconnected by tubes, the air sacs deliver oxygen to almost every part of the body—even into the hollow cavities of the long bones. The air sacs also contribute to the buoyancy of the bird.

Legs
Probably the most effective shock-absorbing mechanism found in nature, each leg consists of three rigid bones with joints that work in opposite directions to cushion the landing.

surfaces such as the elevator and rudder, and on proposed configurations for airplane wings. Then in 1853, after a lifetime of cautious experimentation, Cayley built the world's first successful glider. Crewed by his reluctant coachman, the cloth-winged apparatus was reportedly seen wafting across a valley on Cayley's Yorkshire estate.

Four years later, French sea captain Jean-Marie Le Bris, who studied albatrosses while sailing around Cape Horn, constructed a 46-foot glider closely resembling the majestic seabird. He equipped it with an adjustable, pedal-operated tail and, as an artistic flourish, a long artificial beak. Racing downhill atop a horse-drawn cart, Le Bris apparently gained enough windspeed to ride his *Albatross* to a height of 300 feet. Unfortunately, he crashed on a second flight—shattering both his whimsical glider and his leg.

Such mishaps were all too frequent in flying's early days. Even the greatest of the glider pioneers, the German engineer Otto Lilienthal, who in six years of intensive experimentation made nearly 2,000 flights, took his share of tumbles, the last one fatal. Lilienthal's legacy was vital, however, for it inspired Orville and Wilbur Wright, joint owners of a bicycle repair shop in Dayton, Ohio, to turn their mechanical skills to aviation. The Wrights' first goal was to develop an efficient control system that would help prevent the kind of crashes that had killed Lilienthal. The brothers became dedicated bird-watchers, and it was Wilbur's observation of turkey vultures that provided the necessary clue. He noted that, when tipped sideways by a gust of wind, the bird righted itself by a slight twisting of its wing tips; the leading edge of one wing tip would turn down, while the other turned up. Wilbur reasoned that a system enabling pilots to twist, or warp, a glider's wings in this manner would provide adequate control. After successfully testing the theory on a kite, the brothers incorporated wing warping into their first glider, built in 1900.

By 1902 they had made nearly 1,000 flights over the sand dunes at Kitty Hawk, North Carolina, a site they had chosen for its rolling terrain and steady wind. They were now experimenting with their third glider, and had perfected a flight-control system that included an elevator to regulate the glider's nose position, and a rudder to help control the tail. To measure the performance of various wing designs, they had even constructed a primitive wind tunnel—an open-ended wooden box with a glass window and a steel fan driven by a one-horsepower gasoline engine—that forced air past model wings. Soon they had a stable, maneuverable aircraft. To achieve powered flight, all they needed was a lightweight gasoline engine. Finding none that suited them, the remarkable brothers designed their own, then carved its twin wooden propellers.

On December 17, 1903, the Wrights' *Flyer* made its wobbly maiden voyage—120 feet across the sand dunes of Kitty Hawk at shoulder height, with Orville piloting. Twelve glorious seconds of powered flight christened a new era in aviation. The brothers made three more flights that day, achieving in one of them a distance of 852 feet, and staying aloft for 59 seconds. Paying tribute to their feathered teachers, Orville wrote: "Learning the secret of flight from a bird was a good deal like learning the secret of magic from a magician. After you once know the trick, you see things that you did not notice when you did not know exactly what to look for."

As late as 1912, pioneer airmen were still trying to imitate the flapping flight of birds. Here a rare photograph captures an Austrian tailor, Franz Reichelt, as he prepares to leap off the Eiffel Tower. His billowing, parachute-like suit barely slowed his fall, and he plunged 190 feet to his death.

GETTING AN ANGLE ON LIFT

One secret the Wright Brothers explored was the phenomenon of lift—a basic aerodynamic effect that allows an aircraft to overcome the pull of gravity and so stay aloft.

Despite appearances, airplanes do not ride on a cushion of air; the process is more one of suction. The principle was first explained by the Swiss scientist Daniel Bernoulli, who in 1783 discovered that increasing the velocity of a fluid, as it moves through a pipe or other conduit, will diminish the pressure it exerts against the conduit's sides. For example, when a wide river is forced to negotiate a narrow canyon, the water pressure on the riverbed is measurably lower than in the gentler stretches upstream and downstream. A pair of pressure gauges, one placed in the canyon's swirling rapids, the other lodged in a quiet pool well below them, gives dramatic proof of this.

In the world of aeronautics, air is as much a fluid as water, and it behaves in just the same way. The shape of an aircraft's wing, called an airfoil, is specially designed to create lift, the force that makes heavier-than-air flight possible. A wing's upper surface is curved, while its bottom is straight. As the oncoming airstream divides to flow around the wing, it must travel faster over the curved upper surface (which acts much like the canyon) than across the straight bottom surface—since the distance is greater. According to Bernoulli's Principle, the pressure of the upper

A replica of the Wright Brothers' 1902 glider soars over the dunes of Kitty Hawk, North Carolina. By changing the angle of the wing tips—a technique known as wing warping—the Wrights could control their craft in an unsteady wind and bank it from side to side. With the addition of a front elevator to vary the glider's pitch, and a rear rudder for steering, the brothers had invented a trustworthy, controllable aircraft. All that it lacked was a power plant.

surface will be less than the pressure on the bottom surface. As a result of this difference, the wing lifts.

Reinforcing this basic flow pattern is a secondary sequence, which further augments the difference in pressure. As the airstream flows over the wing, it comes off the wing's trailing edge with a spin, known as the starting vortex. A law of aerodynamics states that a vortex always produces a counter vortex of equal strength that rotates in the opposite direction. Underneath the wing, air circulating in the counter vortex clashes with the main under-the-wing airstream. Velocity diminishes, and air pressure below the wing consequently increases. Above the wing, both airflows move in the same direction, and their combined velocity boosts the above-wing airflow. The result is an increase in lift.

Other factors affect the amount of lift a wing generates, from the manipulations of the pilot to the wing's basic design. Simply pushing an airplane's throttle increases lift; as airspeed rises, so does the pressure differential. Another aspect controlled by the pilot is the angle at which the wings meet the oncoming airstream. By inclining the wings at a slight angle—called the angle of attack—the air meets the undersurface of the wing at a steeper angle, and creates an even greater pressure difference above and below the wing. Generally the most efficient angle of attack for airplanes is about four degrees to the oncoming airstream.

The main design factors affecting lift are the wing's surface area and its camber—the degree of curvature of its upper surface. More camber, and more surface area, each produces more lift. As a rule of thumb, aircraft designed to fly at low speeds, like gliders, need extra lift and so carry large, highly cambered wings. Since velocity also substantially increases lift, high-speed jets can get by with smaller, flatter wings.

Aeronautical engineers express lift in pounds per square foot, and the heavier an aircraft the more lift it obviously requires from its wings. A lightweight glider can soar with ease on the miserly two pounds of lift supplied by each square foot of its long, fully cambered wings. A heavier Piper Tomahawk trainer moves at a

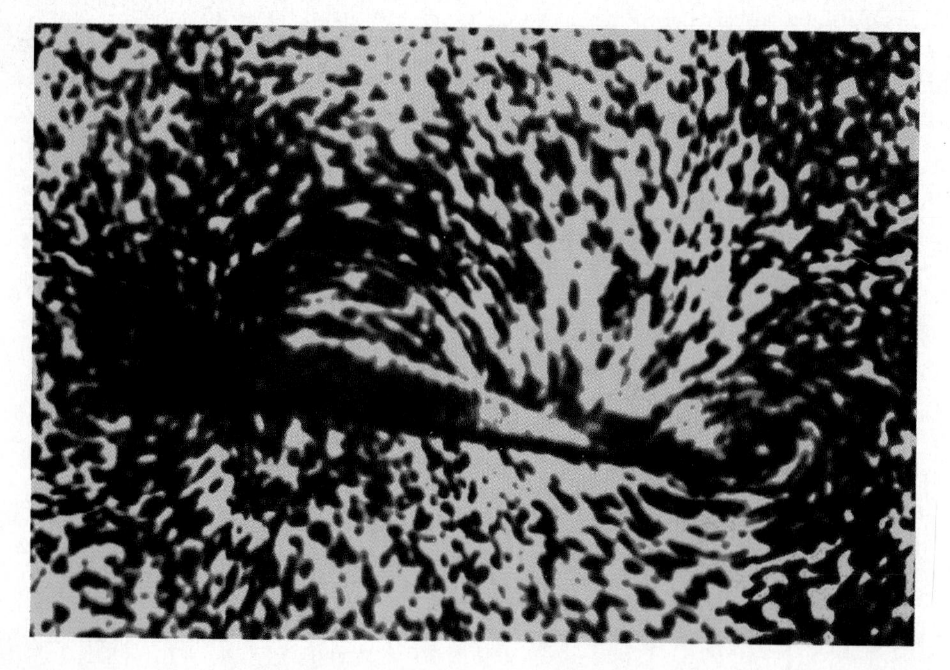

A pattern of oil droplets, sprayed into a windstream moving at about 100 miles per hour, reveals the circulation of a fluid around a stationary airfoil being tested inside a wind tunnel.

faster pace, and so its shorter, less-cambered wings provide all the 13.5 pounds of lift it needs to sustain level flight. The fire-breathing, 50-ton F-14 Tomcat relies on its sensational speed to eke 82 pounds of lift from each square foot of its razor-thin wings.

THE WEIGHT PROBLEM

"Why does fuel have to be so heavy? If gasoline weighed only a pound per gallon instead of six, there'd be no limit to the places one could fly." So wrote Charles Lindbergh about his most troublesome problem as he prepared, in the spring of 1927, to make the world's first nonstop transatlantic flight.

Lindbergh's complaint was understandable. His 2,150-pound monoplane, the *Spirit of St. Louis,* would need to burn an estimated 425 gallons of fuel as it flew the 3,600 miles between New York and Paris. But to sustain this weighty load, given the technology of the day, the plane would need the extra lift of an impossibly large wingspan. So designer Donald Hall took a calculated risk. He gutted the *Spirit* of every nonessential, until it was little more than a flying gasoline tank. This allowed him to cut the wingspan to a manageable 46 feet—and Lindbergh to lift off to aviation glory.

Today's designers face similar dilemmas of weight, lift and structural integrity. In engineering terms, weight is the downside force opposing lift; but unlike lift, it cannot be aerodynamically manipulated. The only solution is to reduce it as much as possible. To do so, designers have turned to a variety of composite materials, light in weight but of great strength, for use in airframe construction.

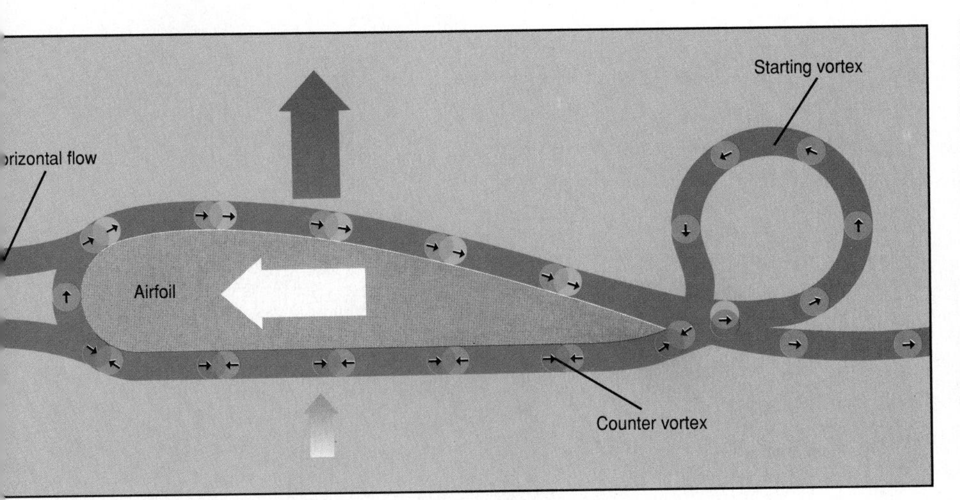

THE SECRET OF LIFT
This cross section of an airplane's wing moving through the air reveals its airfoil shape, and the activity of the air as it sweeps past. As the horizontal airstreams above the wing (yellow) and below it (red) join at the trailing edge, they form a circular eddy, called the starting vortex, and produce a counter vortex that rotates counterclockwise around the airfoil (orange). Underneath the wing, the counter vortex meets the horizontal flow and checks its speed. Above the wing, both airflows move in the same direction, and their combined velocity increases. The result is a substantial difference in pressure above and below the wing that sucks it upward.

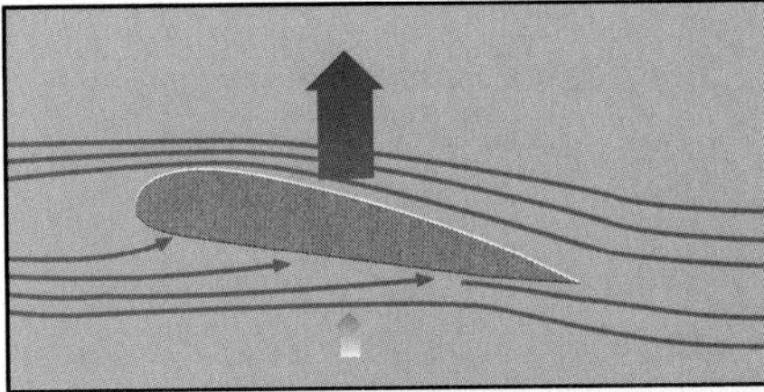

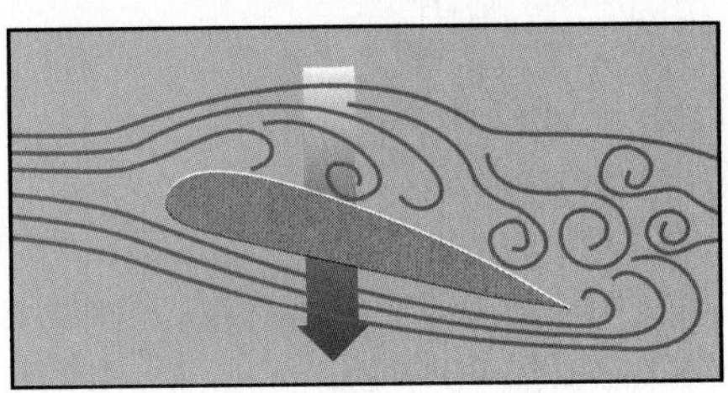

Raising the angle of attack—the angle at which the wings meet the oncoming airstream—increases lift. For example, steepening the angle of attack from 4° to 8° (top) *doubles the total lift force. When the angle of attack reaches about 14°, air flowing over the wing becomes turbulent; the low-pressure area above the wing starts to dissipate and destroys lift* (bottom).

The first airframe material was wood—the Wrights' *Flyer,* for example, used it—and wood itself is a marvelous organic composite. With its long cellulose fibers locked in a rigid matrix of lignum, it is both flexible and sturdy. Wood has its limits, however, and designers soon began using metals such as steel, iron and aluminum. Now, with the physical stresses of today's high-speed aviation, even better materials are needed. So scientists are designing them, molecule by molecule, often mimicking the structure of natural objects. For example, a carbon composite, considerably lighter and six times stiffer than aluminum, draws most of its strength from micro-thin carbon fibers set in an epoxy matrix. And the strength is extraordinary. A filament of carbon or graphite can be spun out to miles in length, made finer than a human hair, and it will not break until the stresses upon it reach up to 20 times the breaking point for metals.

Structure is key in the new materials, and nature provides clues here as well. The interlocking cells of a honeycomb makes it extremely rigid in one direction and resilient in the other. This property is useful in airframes, so sheets of honeycomb made of aluminum, graphite epoxy, and even paper impregnated with plastic, are now being used in airliners and military jets.

Besides strength and lightness, another sought-after quality is heat resistance. Often, it is provided by metals—superalloys—blended of exotics such as tungsten, columbium, lithium and titanium. Exceptionally strong at high temperatures, superalloys are now being used in the compressors of jet engines, where temperatures can reach up to 2,500° F. They also find application in supersonic aircraft, where aerodynamic heating caused by high speeds can expose an aircraft's surface to temperatures as high as 1,000° F. Ceramics, capable of withstanding even greater heat than metals, may soon find an application in aero engine parts. And lightweight ceramic-reinforced polymers are already being tested as a material for the flight surfaces of a plane.

In 1986, composite materials were used to build an airplane that set the ultimate distance record. Two American pilots, Dick Rutan and Jeana Yeager, accomplished a nonstop voyage around the world in a specially designed craft called the *Voyager.* As with Lindbergh's *Spirit of St. Louis,* the design yardstick for *Voyager* was its fuel load. Designer Burt Rutan, Dick's brother, calculated that an ordinary aluminum-frame plane, to contain the fuel it would need for a round-the-world flight, would have to be as big as an aircraft carrier. His solution was to build the plane almost entirely of composites. He glued together layers of carbon-fiber cloth and honeycombed paper, sandwiched between a tough skin of graphite fibers, creating a material seven times stronger than aluminum. The result was a 2,860-pound aircraft—just 100 pounds heavier than Lindbergh's—capable of carrying

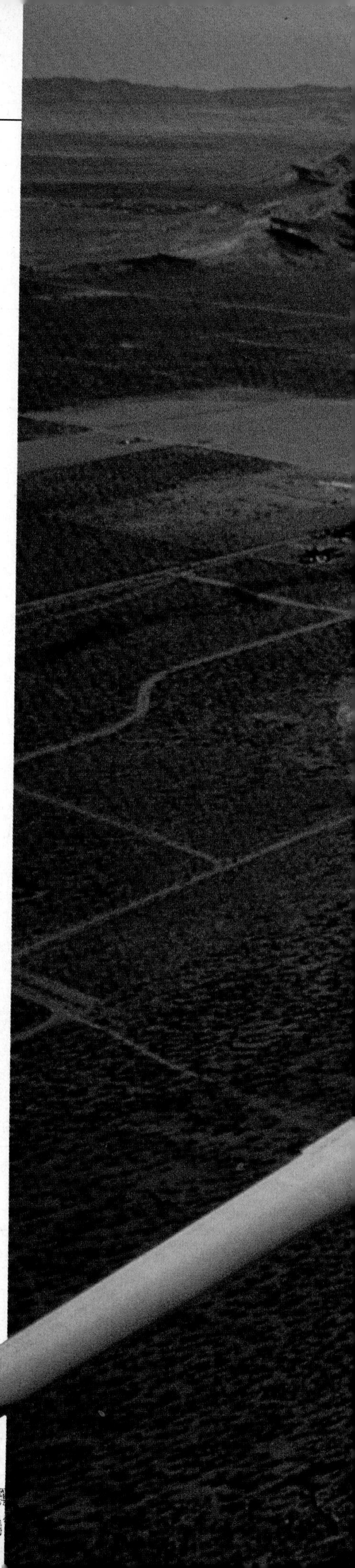

The lightweight, spider-slim Voyager *ghosts across the Mojave desert en route to its record-breaking, 25,000-mile nonstop circumnavigation of the globe. Pilots Dick Rutan and Jeana Yeager—Jeana flying and Dick napping—huddle in* Voyager*'s tiny unpressurized cabin. Measuring just 3 1/2 by 7 1/2 feet, it was little larger than a phone booth; shifting places without knocking the controls was a minor feat of acrobatics.*

two passengers 25,000 miles nonstop, without in-air refueling. A consumate fuel miser, *Voyager* averaged 36 miles to the gallon.

Weight still remains an aircraft designer's toughest adversary. Jumbo jet or Mach 2 fighter, the airframe must be light and sturdy. As an industry saying puts it, an aeronautical engineer is a man "who must build for one pound of weight what any fool could do for two."

THE STRUGGLE FOR STABILITY

An experienced jet test pilot who recently flew a replica of Louis Blériot's famed 1909 monoplane, the first heavier-than-air machine to cross the English Channel, was amazed that the French inventor managed to fly at all. He found that the mental concentration, and continuous control adjustments required just to stay on an even keel, were unspeakably difficult and tiring. What most airplanes of the Blériot era lacked was stability.

Stability is the ability of an airplane to remain in the flight attitude the pilot selects, and to self-correct when it is upset by air turbulence. Slight air disturbances, such as updrafts or gusts of wind, tend to send it out of kilter in any of three dimensions, as shown in the diagrams at right. So aviation designers soon developed control surfaces to help preserve equilibrium: tail fins to keep the aircraft from weaving back and forth, horizontal stabilizers to discourage pitching, and wing alignments that prevent it from rolling from side to side.

DESIGNS FOR STABILIZING FLIGHT

An object traveling through the air is vulnerable to movement along its horizontal, vertical and longitudinal axes. The airplane pictured here displays the built-in design characteristics that automatically correct for movement along these axes.

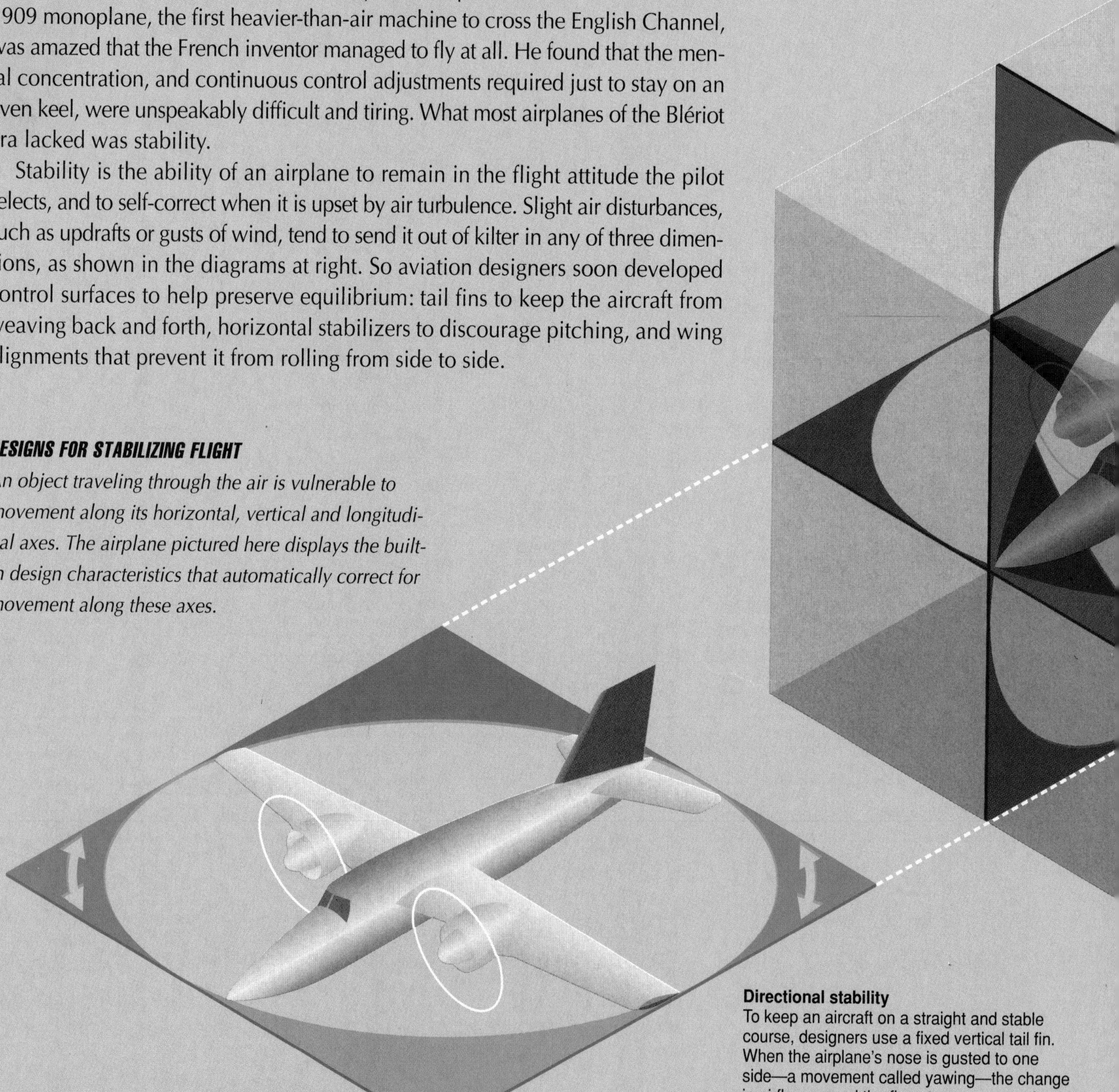

Directional stability
To keep an aircraft on a straight and stable course, designers use a fixed vertical tail fin. When the airplane's nose is gusted to one side—a movement called yawing—the change in airflow around the fin sets up an opposite, correcting force. Like the tail of a weathercock, it pulls the nose back into line with the airflow.

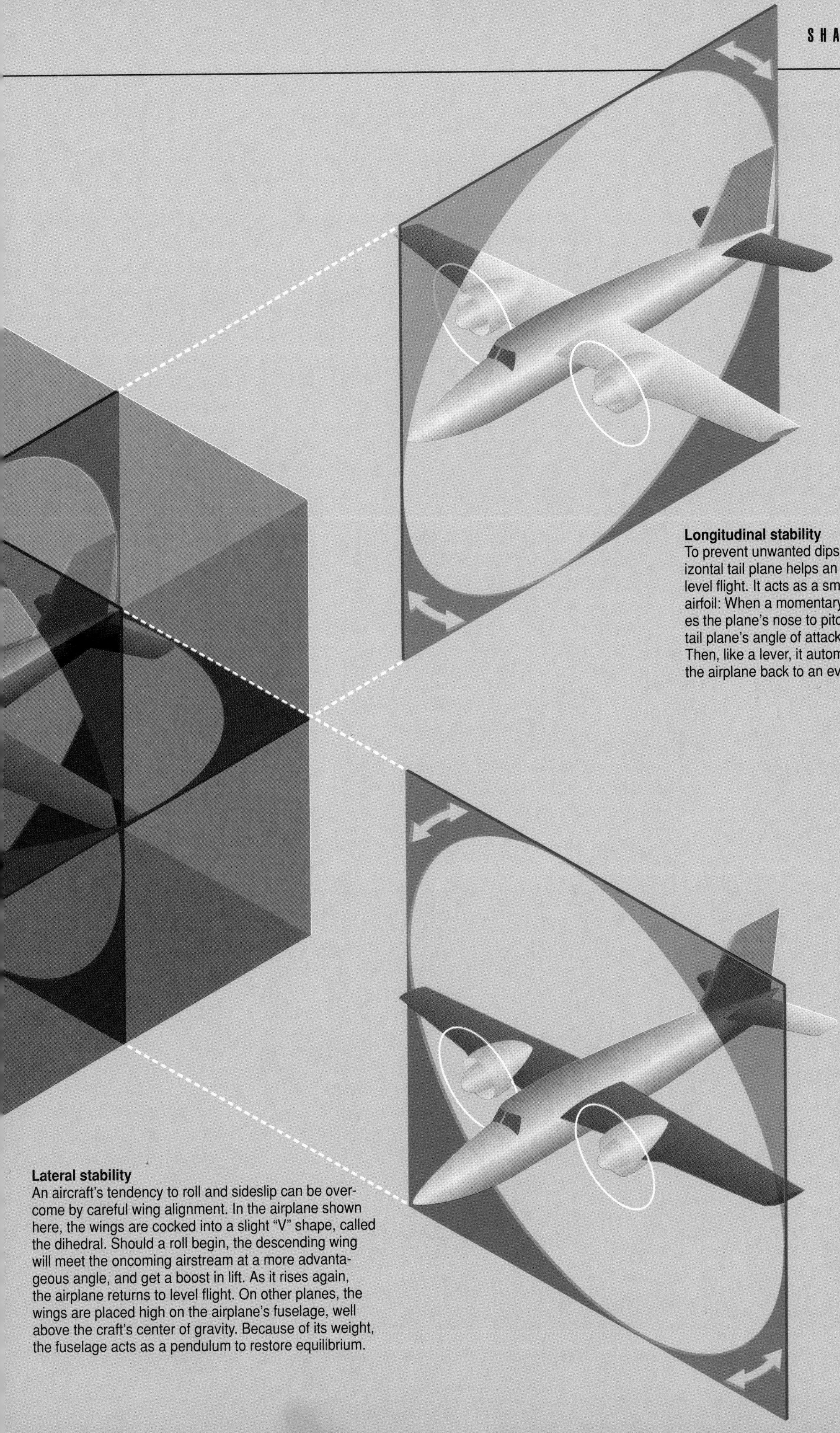

Longitudinal stability
To prevent unwanted dips or nose-ups, a horizontal tail plane helps an airplane maintain level flight. It acts as a small supplementary airfoil: When a momentary disturbance causes the plane's nose to pitch up or down, the tail plane's angle of attack also changes. Then, like a lever, it automatically hoists the airplane back to an even keel.

Lateral stability
An aircraft's tendency to roll and sideslip can be overcome by careful wing alignment. In the airplane shown here, the wings are cocked into a slight "V" shape, called the dihedral. Should a roll begin, the descending wing will meet the oncoming airstream at a more advantageous angle, and get a boost in lift. As it rises again, the airplane returns to level flight. On other planes, the wings are placed high on the airplane's fuselage, well above the craft's center of gravity. Because of its weight, the fuselage acts as a pendulum to restore equilibrium.

Testing Designs for Flight

"Test first, fly later": This has always been the order of business in the history of aviation research. The wind tunnel is the primary testing tool of aeronautical engineers. Its invention dates to 1871 when Francis Wenham, a marine engineer fascinated by the idea of powered flight, designed the first crude test chamber. He used a steam-powered fan to force air past the model of a wing design.

A modern wind tunnel employs the same principles as its 19th-Century predecessor but it yields far more accurate results. Inside the tubelike structure, a carefully controlled airsteam, usually produced by a huge fan, flows over scale models of airplanes or their components in the tunnel's test section. Computers, connected to the suspended model, measure and record airflow and aerodynamic forces.

Tests are conducted with models by relying on a formula, developed in 1883 by British physicist Osborne Reynolds. This formula takes into account the size and shape of the proposed design and the characteristics of the air flowing over it (its speed, density and viscosity). The equation results in a reynolds number. Next a scale model is tested in a wind tunnel, under conditions that yield the same reynolds number and the model's performance dictates the success or failure of the design.

A small, slow-speed plane such as a Cessna Skymaster has a relatively low reynolds number that is easy to simulate in a wind tunnel. But testing models of large, high-speed planes poses problems. There are two solutions—larger tunnels and larger models or manipulation of the properties of the air flowing over the model. Air in the tunnel can be pressurized, substituted with a denser gas such as nitrogen, or cooled to subzero temperatures to increase viscosity.

An electronic version of a wind tunnel is emerging as a useful tool of aerodynamicists. Called Computational Fluid Dynamics (CFD), it uses high-speed computers to generate mathematically the flow of fluid over a computer-designed model. The computer can analyze the aerodynamic forces on the aircraft's surfaces, quickly sort through a large number of possible design modifications and present the best solution.

Energy conservation is one of the main goals of wind-tunnel testing and CFD research. Commercial airliners in the United States consume more than ten billion gallons of fuel every year; even a one-percent improvement in fuel efficiency would save millions of gallons.

Computational fluid dynamics provides aerodynamicists with a time- and money-saving testing tool. Shown below, the computer-generated wind flow onscreen appears as multicolored trace lines over an F-16's wings and fuselage. The colors represent air pressure; red depicts the high-pressure areas, and blue the low.

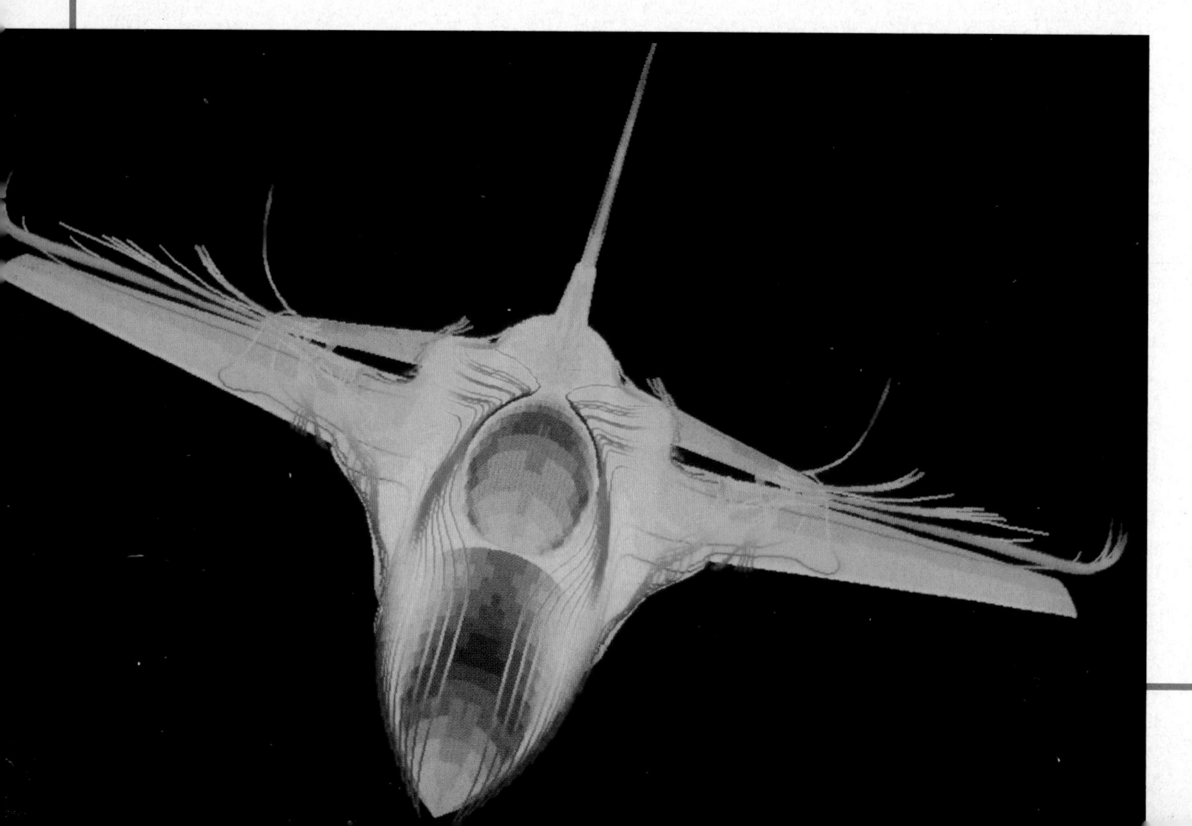

The National Transonic Facility (right) *is a new kind of wind tunnel that uses cryogenic or extremely cold nitrogen gas at high pressure to test small-scale models of advanced aircraft or spacecraft as they fly through the sound barrier. Cryogenic temperatures as low as minus 300° F. reduce the viscosity and increase the density of the fluid flowing through the tunnel.*

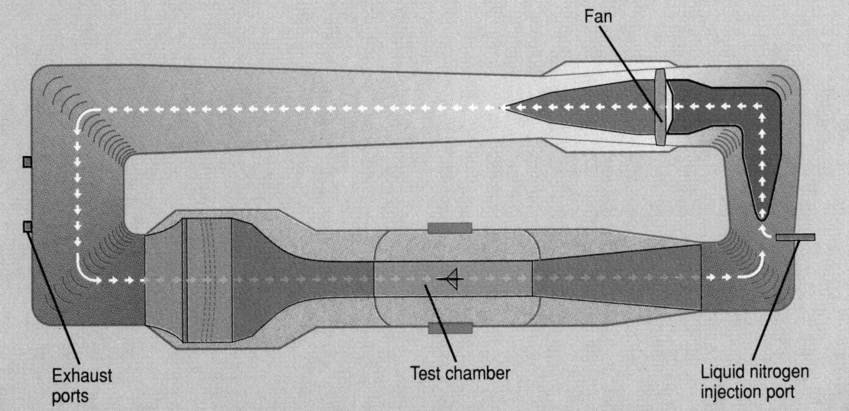

A technician inspects the fan blades of the National Transonic Facility at the Langley Research Center in Virginia. Electric motors, capable of generating 126,000 horsepower, drive the fan blades that circulate air or supercold nitrogen around the wind-tunnel circuit at speeds of more than Mach 1.

Along with stability comes balance, for unless an airplane's weight is properly distributed it will soon tumble out of the sky. Aircraft fly best when their center of gravity is located just ahead of the point where the three dimensions of stability intersect. This means that whenever an airplane is made ready for takeoff, care must be taken that fuel and payload are correctly placed—like trying to position children of different weights on a seesaw, but considerably more complicated. Computers are now replacing the graphs and calculators commonly used by pilots to perform their vital preflight loading calculations. Some modern airliners, such as the supersonic Concorde, have onboard computers that feed information automatically to a center-of-gravity indicator in the cockpit.

THE POWER PLANT

Even while the early glider pioneers were searching for a flyable airframe, a number of farseeing airmen were attempting to solve the next problem: power. All kinds of bizarre power plants have been tested over the years, driven by everything from carbonic acid to gunpowder to elastic bands. One early favorite was steam, and in 1894 Britain's Sir Hiram Maxim, inventor of the machine gun, used a pair of 180-horsepower compound steam engines to lift his galumphing four-ton *Leviathan* momentarily from its launching rails. What was needed for true powered flight—beyond a suitable airfoil—was an engine that could deliver generous amounts of horsepower for relatively little weight. Unless an engine can move its airplane forward at sufficient speed, it will not create the lift required to raise the airplane, the payload, and the engine itself against the pull of gravity.

The earliest airborne power plants met none of these requirements. Heavy steam-driven devices of ponderous inefficiency, they were carried aloft on balloons and dirigibles—which, being lighter than air, had no need of aerodynamic lift. The first such machine to make the ascent was a 100-pound steam-puffing engine that French engineer Henri Giffard installed on an 88,000-cubic-foot airship in 1852. Weighing 117 pounds for every unit of horsepower it developed, Giffard's engine could barely push itself against a gentle breeze.

A leaner and more muscular type of power was clearly needed. The inspiration came from the automobile, where gasoline-burning, piston-driven, internal combustion engines were proving increasingly effective.

The Wright Brothers used what was basically a modified automobile engine to propel their *Flyer* on its historic leap at Kitty Hawk. Their mechanic, Charlie Taylor, built a four-cylinder gasoline engine that weighed a mere 180 pounds, including its flywheel, and generated about 13 horsepower. It gave the 605-pound *Flyer* just enough lift to become airborne.

Besides building an engine, the Wrights also designed and constructed their own propellers—necessary to convert the power of the engine into propulsive thrust. Propellers of a sort were already being used on the early airships; but these were huge paddles, often more than 20 feet in diameter, that turned lazily to push the balloon-like craft along at a walking pace. The Wrights needed a propeller that would withstand the stresses of spinning at 350 revolutions per minute—a rate reflecting their vision that a propeller should not be a paddle, but an airfoil—literally, a wing spinning through the air to generate lift in a forward direction, just as the airfoil shape of a fixed wing creates upward lift. As a propeller rotates, air

A PORSCHE THAT FLIES

This cutaway view of a state-of-the-art Porsche PFM 3200 reveals the essentials of propeller-driven thrust. The airborne equivalent of the Porsche 911 sportscar engine, this aero engine is the power plant of choice for sporty private aircraft such as the Mooney.

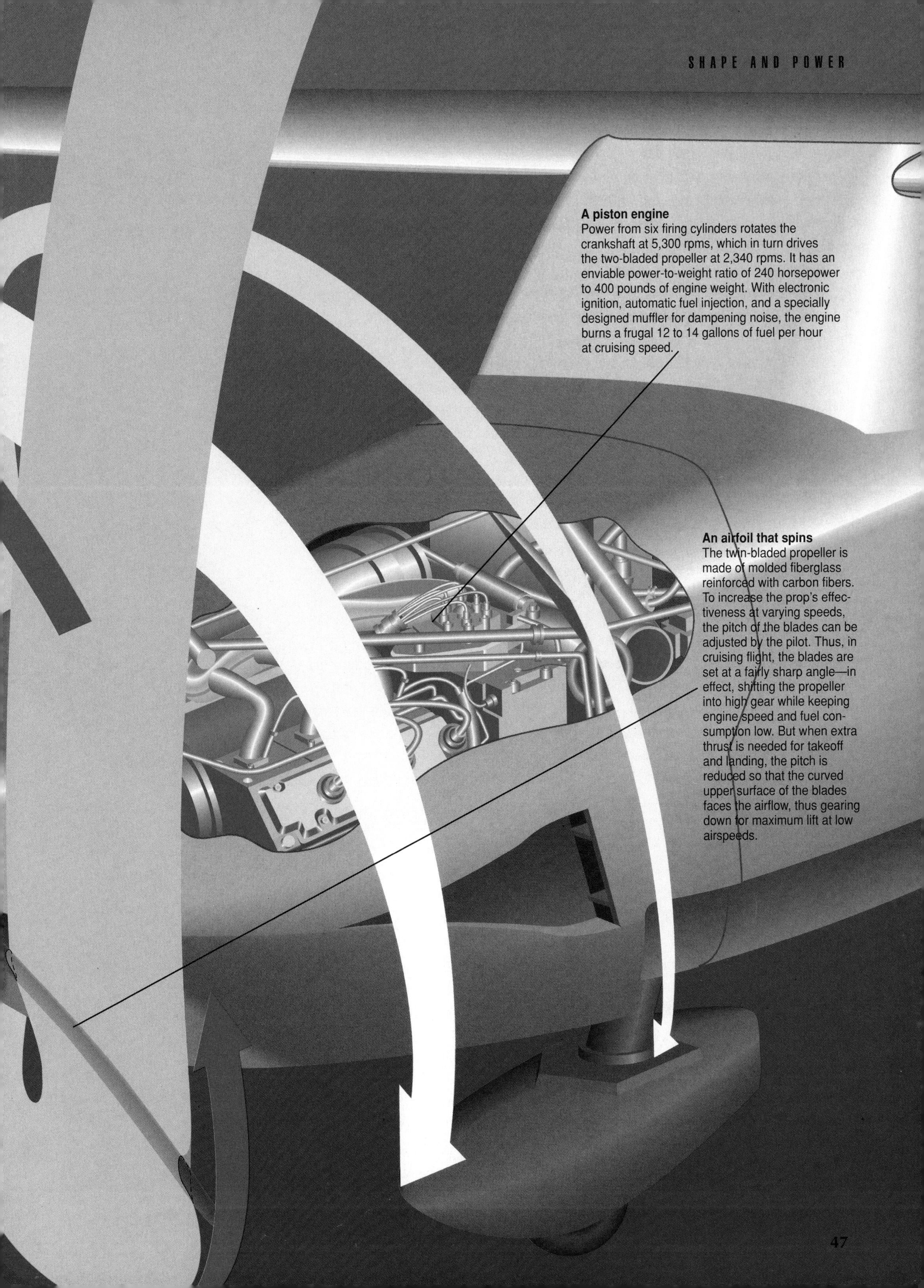

A piston engine
Power from six firing cylinders rotates the crankshaft at 5,300 rpms, which in turn drives the two-bladed propeller at 2,340 rpms. It has an enviable power-to-weight ratio of 240 horsepower to 400 pounds of engine weight. With electronic ignition, automatic fuel injection, and a specially designed muffler for dampening noise, the engine burns a frugal 12 to 14 gallons of fuel per hour at cruising speed.

An airfoil that spins
The twin-bladed propeller is made of molded fiberglass reinforced with carbon fibers. To increase the prop's effectiveness at varying speeds, the pitch of the blades can be adjusted by the pilot. Thus, in cruising flight, the blades are set at a fairly sharp angle—in effect, shifting the propeller into high gear while keeping engine speed and fuel consumption low. But when extra thrust is needed for takeoff and landing, the pitch is reduced so that the curved upper surface of the blades faces the airflow, thus gearing down for maximum lift at low airspeeds.

flows around the blade, moving faster over the curves of the blade's leading edge. The motion lowers the air pressure in front of the blade and pulls the aircraft forward. Figuring these forces in painstaking tests and calculations, the Wrights designed and handcarved a pair of elegant propellers from blocks of laminated spruce. They linked them to their engine with a system of chains and sprockets much like those on a bicycle.

During the next five years, the Wrights boosted their engines to 30 horsepower. But by then Europe had taken the lead in engine technology. The great leap forward came in 1905, when France's Léon Levasseur, reworking a speedboat engine, devised a V-8 power plant that delivered 50 horsepower and weighed only 110 pounds. Its remarkable power-to-weight ratio—one horsepower for each 2.2 pounds of engine weight—would not be exceeded for 20 years. But Levasseur's engine had problems with its fuel injection and cooling systems, and frequently broke down. It never achieved the wide commercial success of another innovative French design—the legendary Gnome rotary.

The Gnome was one of the first production engines manufactured exclusively for airplanes. The first model appeared in 1907, and its design was truly revolutionary. Five cylinders, rated at 10 horsepower each, were placed like wheel spokes around a fixed crankshaft. When the cylinders fired, the entire engine would turn, imparting its spin to a propeller bolted directly to it. Because of this rotary action, the engine was self-cooling; thin metal fins on each cylinder allowed the heat of combustion to dissipate directly into the passing air. It also acted as its own flywheel, with a marked savings in weight. The Gnome was not perfect: It was a glutton for fuel, and the mass of whirling cylinders tended to make the aircraft attached to it difficult to control. It also sprayed lubricating fluid—castor oil—back over the pilot. But the Gnome was light, peppy, smooth-running and reliable, and airplane designers loved them. By 1916, almost 80 percent of all aircraft ran either on Gnome rotaries or on engines derived from them.

THE FORCE WITHIN A TOY BALLOON

Inflated and tied shut, a toy balloon is a miniature engine waiting to start up. The compressed air within it pushes equally on all sides; but until the air finds an outlet, the balloon remains stationary.

This was the decade of World War I, when aircraft were first enlisted as instruments of battle, and designers turned every effort to improve both power and performance. Car manufacturers—Rolls-Royce, Mercedes and Hispano-Suiza—produced aerial power plants of ever-increasing horsepower. The rotary design gave way to other configurations, where the cylinders were fixed and the crankshaft turned. The result was a welcome cut in fuel and oil consumption, and greater efficiency on long-range flights. By 1918, water-cooled engines with banks of 12 cylinders were developing 300 to 400 horsepower. Nor did the rate of innovation decrease with the Armistice. The 1920s saw such refinements as variable-pitch propellers that effectively gave airplanes gearshifts; improved carburetors that could be adjusted for various altitudes and power settings; and starter motors that removed the dangerous task of hand-swinging the prop before each flight.

The United States became world leader in engine design following Charles Lindbergh's epoch-making Atlantic flight. The 500-pound, 220-horsepower air-cooled Wright Whirlwind that powered the *Spirit of St. Louis* set new standards of power for weight, economy and reliability. The Whirlwind was an air-cooled radial, with stationary cylinders placed in a collar around a spinning crankshaft. Its main drawback was wind resistance from the protruding cylinders, but this was solved by covering the engine with a streamlined cowling. Within ten years, Wright and the

CONTINUING PROPULSION
If a means could be found to maintain pressure by pumping in more air, the balloon would keep moving—like a jet supplied with a steady flow of fuel.

THE FORCE IN ACTION
When the balloon's neck is opened, the trapped air rushes out. As the air escapes, it upsets the balance of forces inside the balloon, causing an equal and opposite reaction that propels the balloon forward like a squid in water—or a jet through the sky.

rival firm of Pratt & Whitney were building radial engines producing nearly 2,000 horsepower. Some of them are still in use today.

By the end of World War II, monstrous piston engines sprouting 28 cylinders were capable of developing 3,500 horsepower. Even larger engines were on the drawing boards, as airline companies prepared for a postwar boom in commercial aviation. Yet the days of roaring pistons and flashing propellers were numbered. In the last months of the war, a new phenomenon streaked across the sky: the jet.

Both sides in the conflict had been working to perfect this device. For well over a decade, a few visionary engineers had been toying with the idea that a stream of hot exhaust from a gas turbine might be used to drive an airplane. The first practical test of such a turbine, in a machine shop in 1937, almost ended in disaster. Recalled its British designer, Frank Whittle: "With a rising scream the engine began to accelerate out of control. Everyone took to their heels. I was paralyzed to the spot." Luckily for Whittle, the engine did not blow up. But problems continued to plague designers. Then, in the spring of 1944, the first fully operational jet aircraft, Germany's Messerschmitt 262 thundered into combat at 540 miles per hour—70 miles per hour faster than its swiftest propeller-driven opponent. Though its appearance came too late to alter the course of the war, the Me-262 swept jet technology into the modern age.

The action of a jet engine could not be more simple or direct. Air swallowed in through the front is compressed by whirling fans, mixed with kerosene-like fuel, and ignited. The gases of combustion expand with tremendous force and rush out the jet's tailpipe at speeds approaching 1,700 feet per second. The force of reaction thrusts the engine, and the airplane, forward.

The principle of reaction is one of the oldest in mechanics. It was demonstrated as long ago as the first century A.D., when Hero of Alexandria used it to operate the earliest-known jet engine—a bronze globe spun by jets of steam. Isaac Newton set it down as a basic law of physics in the 17th Century: For every action there is an equal and opposite reaction. One example is an inflated toy balloon racing across a room, a jet of air rushing out its open neck and creating a reaction force that propels the balloon forward.

In the natural world, the squid uses the same principle for quick movement in water. Water is drawn into the body, or mantle, where it is stored until needed. When the squid wants to move, it squirts the water backward through a funnel-like opening in the mantle's rear end. Octopi use the same technique, and some species of octopus expel the water with such force that they shoot above the surface and glide a short distance in the air.

Jet engines brought a marked improvement in power and thrust. They replaced the pounding pistons and intermittent explosions of conventional power plants with smooth-spinning turbines and steady combustion. The result was far more horsepower for each pound of engine weight. Whereas piston engines typically produced one-half pound of thrust for a pound of engine weight, today's jumbo jet engine produces four pounds of thrust for each pound of engine weight. Even so, the early jet engines were spendthrifts of fuel, especially at low speeds, and so were better suited to the performance demands of military flying.

Cost-conscious commercial airlines compromised by using turboprops, in which the superior power of a jet turbine is used to drive a conventional propeller. But while this system saved fuel and gave pilots greater control during takeoff, landing, and other slow-speed maneuvers, it had dire limitations. The main problem was the propeller itself. At higher throttle settings, the propellers would start spinning so fast that their tips would approach the velocity of sound. The result was a dangerous increase in air turbulence. In fact, at about 450 miles per hour, the efficiency of the propellers dropped drastically and severe vibrations set in due to the enormous increase in drag.

Designers soon took another tack, focusing on jet exhaust. Various tests and calculations showed that the high-speed exhaust from the early turbojets was inherently inefficient; much more thrust can be obtained from a larger stream of slower-moving air. So in the early 1960s engineers began searching for ways to simultaneously slow down the jetstream and expand its volume. Large fans were installed in front of the jets' compressors, supplying extra air that bypassed the combustion chamber. This supplementary airstream, when mixed with the hot exhaust gases, slows them down and provides the sought-after extra boost of thrust.

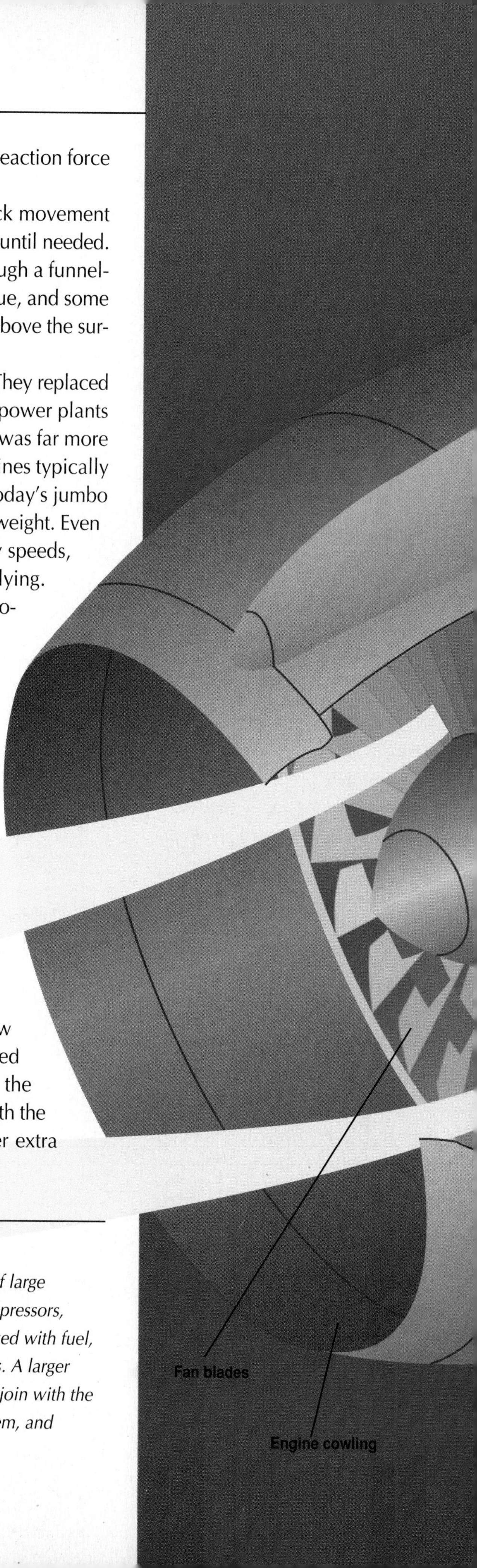

ANATOMY OF A ROLLS-ROYCE

This Rolls-Royce RB211 turbofan displays its jet-powered efficiency. A pair of large whirling fans sucks air into the engine. Some of the air is directed to the compressors, which force it under tremendous pressure into the combustion chamber; mixed with fuel, ignited and expelled, it drives the turbines that turn the fans and compressors. A larger stream of air, pushed back by the fan blades, moves through bypass ducts to join with the exhaust gases as they spurt through the tail cone—cooling them, quieting them, and delivering most of the thrust.

Heated air

Annular combusters

ressors

Exhaust

Tail cone

Turbines

Combustion chamber

Bypass air

Engine core

The turbojet
Air drawn in through an intake is compressed by spinning fan blades and forced into a combustion chamber. Fuel is injected, and the mixture ignites in a continuous explosion. The expanding gases exit the chamber at about 2,300° F. and twice the speed of sound, providing a powerful forward thrust.

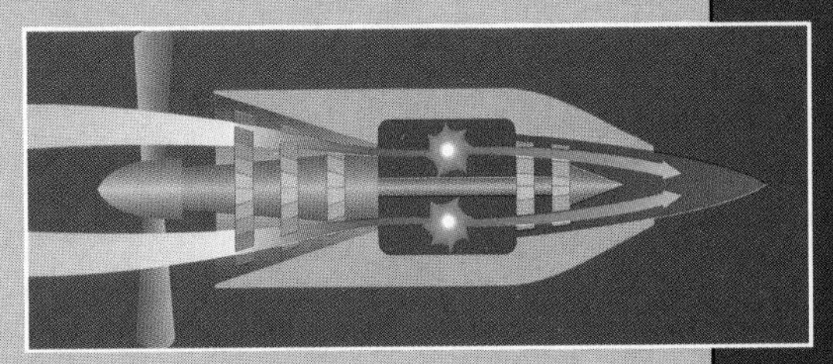

The turboprop
A turboprop unites the power of a jet with the low-speed efficiency of a propeller. Turbines, located behind the compressor turbine, crank the propeller through a system of gears, which slow down its speed. The pull of the propeller provides 90 percent of the thrust; the jet exhaust supplies 10 percent.

Pedal Power

"No bird soars too high if he soars with his own wings," wrote William Blake in 1793. For centuries humans too have strived to fly under their own power. The dream has finally been realized in the form of pedal-powered flying machines, designed with the help of computers and incorporating the latest in synthetic materials.

Shown here is *Daedalus '88*, the culmination of three years of research into human-powered flight. In this remarkable creation, the feathers and wax wings of legend have given way to polyester and graphite. Everything was screened for weight—even the glue. The result was a fragile but high-tech craft that flew more than three times the distance of previous world records for human-powered flight.

Although just about anyone can fly one of these winged bicycles, the energy and power output varies according to the condition of the pilot. Thus endurance and test flights are powered by championship cyclists.

Gearboxes transform the cyclist's energy into mechanical power to spin the propeller. To maintain a cruising speed of about 18 miles per hour, the pilot must pedal at a rate of 75 rpms. To find a comfortable pedaling rate, the pilot adjusts the pitch of the blades much like a cyclist switches gears. Steering demands exacting hand-and-foot coordination. The pilot must continue to pedal while simultaneously operating the rudder and elevator controls with his hands. Coming in for a landing, he slows down and the craft glides to a soft touchdown.

Tailplane
Similar in structure and design to the wings, its spars are made of carbon fibers; the ribs are fashioned from Styrofoam.

Boom
A 29-foot boom supports the tail and the twin-bladed propeller.

Wing ribs
Each of the 102 ribs is cut from 1/4-inch-thick Styrofoam, encased in basswood and reinforced with balsa wood.

Wing spars
Hollow aluminum tubes made from 12 layers of superlight, superstiff graphite epoxy fibers. The walls of the spars are about as thick as a dime.

Covering skin
Resembles cellophane, but made of sheets of transparent polyester.

Propeller pitch cable
Adjusts the pitch of the propeller to regulate the bite of air that it takes during flight: low pitch for high rpms and power on takeoff, and higher pitch for endurance flying and cruising.

Cockpit
The pilot sits on a piece of nylon webbing slung between an adjustable seat frame; an opening in the roof allows fresh air to circulate. An altimeter and a speedometer are the only dials in the cockpit.

Propeller
Constructed of foam, carbon and plastic fiber, the propeller is shaft-driven and connected to the pedals via two gearboxes.

Control stick
Enables the pilot to maneuver the rudder and elevator.

Pedals
Titanium spindles with a plastic roller sheath; each revolution of the pedals results in one-and-a-half revolutions of the propeller.

The sun rises over Daedalus '88 *as it takes wing across the Aegean, following the route of its namesake. Greek cycling champion, Kanellos Kanellopoulos, piloted the aircraft more than 70 miles from Crete to the Isle of Santorini—three times the distance of the previous world record for human-powered flight. As he prepared to touch down, a sudden gust of wind shattered the tail boom, and the craft crash-landed just ten yards short of its destination.*

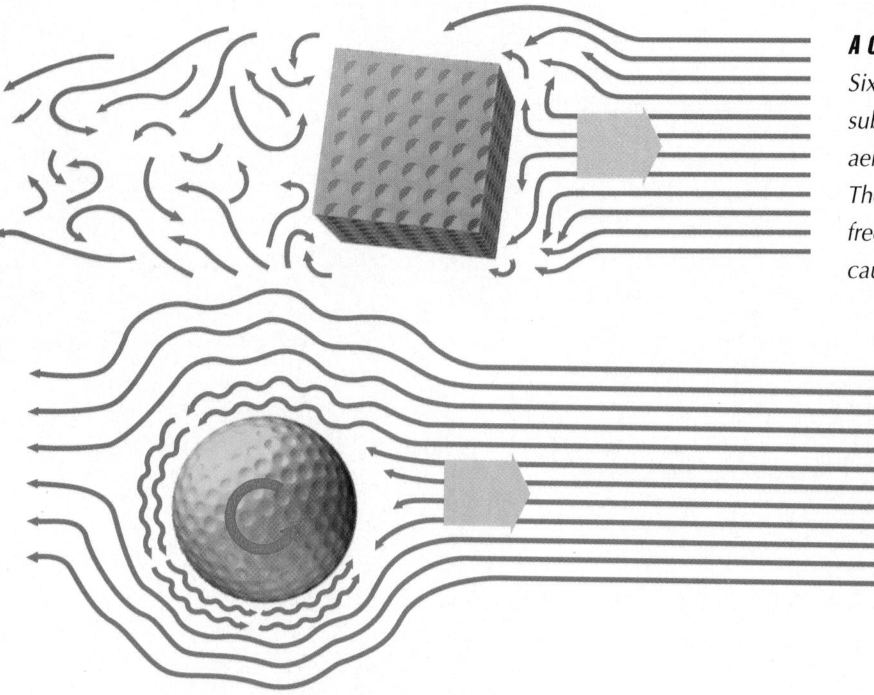

A CLUNKER OF A CUBE

Six flat sides and squared-off corners subject this hypothetical golf ball to aerodynamic drag at its most extreme. The ball's cube shape obstructs the free flow of the passing airstream, causing a turbulent chaos downwind.

A SMOOTH-SPINNING SPHERE

Rounding out the ball's corners allows the airstream to flow past with only minor disruption. In addition, as the ball spins, it generates a degree of lift that increases its trajectory. The effect is further enhanced by the dimples on the ball's surface, which trap a layer of air and set it moving in the direction of spin. The trapped air on top moves with the airstream, while the air below collides with it—much like the counter vortex that helps create lift in an airplane's wing.

So was born the turbofan, the workhorse of modern aviation. Twenty-five percent more fuel efficient than the original turbojet power plants, the turbofan offers better range and more efficient propulsion. It is also quieter and smoother-running, and is capable of delivering effective thrust at both high and low airspeeds. On takeoff—the moment of greatest stress—a Rolls-Royce turbofan engine delivers the equivalent power of 228 Ferrari V-8 engines. In the time it takes to reach a cruising altitude of 35,000 feet, it gulps enough air to fill the *Graf Zeppelin*, more air than a normal man breathes in 20 years.

GRAPPLING WITH DRAG

By 1910 aviators had come to recognize the debilitating effects of drag—the retarding force opposing thrust. Spurred by fortunes in prize money, they strapped bigger and bigger engines to their rudimentary airplanes in search of a winning edge. But tripling the power of their whirling rotaries produced only marginal increases in speed. Their airframes had hit what seemed to be an impenetrable barrier of wind resistance. Here was drag of the most obvious type, and one that thrives on acceleration. Since this kind of drag increases at the square of the airspeed, doubling the velocity of their Blériots to a paltry 68 miles per hour meant a fourfold increase in the forces impeding them.

Moving bodies create drag. Joggers, cyclists and swimmers, boats, trucks and airplanes—all experience a resistance to forward motion through air or water. The bigger they are and the faster they go, the greater the retarding force. There are various ways to mitigate this effect, however. One is to streamline the shape of the moving body.

Streamlining reduces what scientists call form drag—the resistance caused by sharp angles and bulky protrusions. The art of streamlining is centuries old. Greek and Polynesian war canoes sped into battle with their slim prows raising barely a ripple. Today, bullet-shaped racing cars, Olympic bobsleds, even the wind-slicing helmets worn by sprint cyclists, slip through the air with little disturbance. Streamlining in aircraft design produces spectacular improvements. A well-shaped wing strut ten inches thick, and moving through the air at 200 miles per hour, produces less drag than a one-inch steel cable.

Another type of drag is skin friction, the resistance that develops in the thin layer of airstream that passes closest to the airplane. The smoother and more polished the craft's surface, the more easily the air flows past it, and the smaller the effect of skin friction. At low speeds, it plays a minor role in any case. But it jumps dramatically as velocity increases, and in high-speed jets a glassy-smooth surface becomes an essential part of overall design.

A third variety of drag applies only to aircraft. Whenever an airstream flows over a wing or other airfoil, creating lift, some of the energy is diverted to the rear. This has to do partly with the angle at which the wing strikes the airstream, and partly with forces of turbulence that come into play. In either case, induced drag, as it is called, tends to pull the plane backward. It is the price a pilot pays for lift.

Some induced drag is produced by the vortex that forms behind the wing's trailing edge. Another type of vortex also comes into play, and it occurs because air from the high-pressure area beneath the wing tends to leak around the wing's edges toward the low-pressure area above. Most of this seepage takes place at the wing tips, because the wings are slightly swept back. The result is a pair of powerful spiral vortices trailing from each wing tip.

Induced drag is greatest at low speeds and high angles of attack, such as during takeoff and landing. When an airliner lifts off the runway, it spews out massive amounts of wing-tip turbulence. Gliders and other slow-flying aircraft are partic-

THE POWER OF A WING-TIP VORTEX

Mini-tornados stream from an airplane's wing tips as eddies of high-pressure air curl up around them. A differing pattern of airflow above and below each wing—angled inward above, and inclined toward the tips below—helps impart the spiral motion. So strong are these trailing vortices that the wake of a large airplane can literally flip a smaller craft on its back.

ularly vulnerable to drag of this type—one reason their wings tend to be long and slender. By extending the wings and tapering their tips, designers are able to minimize the vortex they create. Another technique is to add fin-like extensions, called winglets, which help reduce vortex drag. They stabilize airflow on the wing tips by decreasing the seepage from under the wing.

FORM FOLLOWS FUNCTION

When the Wright Brothers designed their first powered airplane they had only one concern—to fly. It mattered little that Orville's sand-hugging hop covered less distance than the wingspan of a Boeing 747 and that it could have been outpaced by a man on a fast bicycle. Speed, range, height and payload were all secondary considerations.

This is hardly the case in modern aviation. Today's aircraft are each designed with a specific function in mind. Commercial airliners must carry heavy loads of passengers long distances in reasonable comfort. Jet fighters must be fast, stunt planes highly maneuverable. As a result, designers must compromise, trading one quality of performance for another they deem more desirable.

In nature, the compromises have evolved over time. The shape of a bird's wing is closely linked to the life it leads—and faster is not always better. The broad, rounded wings of a grouse or pheasant allow for quick, furious takeoffs and short flights through heavy forest foliage. Ducks, geese and falcons fly fast over long distances, using slow, steady wingbeats; their wing tips are pointed to minimize drag. Seabirds such as gulls and petrels have long, narrow wings for gliding on steady ocean winds, while hawks and vultures have broad wings with slotted tips that allow them to ride the thermal updrafts generated over land.

In aviation, designers looking for speed must generally sacrifice range; maneuverability is achieved at the expense of miles per hour. But ingenious advances in technology allow a number of aircraft to have it both ways. For example, to minimize drag while cruising near the speed of sound, the 450-ton Boeing 747 has relatively small, swept-back wings. Its problems come during takeoff and landing, when this same wing configuration does not easily provide enough lift to raise the 400-passenger payload off a normal-sized runway. The solution is to temporarily alter the wings' shape by extending their surfaces with huge flaps at the leading and trailing edges.

The dart-shaped Concorde—designed for pure speed, but with a sharp reduction in payload—carries only 125 passengers. Its supersonic cruising speeds dictate a narrow body, and thin, delta-shaped wings that cannot be fitted with lift-generating wing flaps. It must land and take off at such high speeds that only a few of the world's runways are lengthy enough to accommodate it.

The ultimate compromise is the "swing-wing," used in the Navy's Grumman F-14 Tomcat. It has sharp, narrow wings that can be extended to increase lift and stability during low-speed takeoffs and landings. Then, as it soars into supersonic flight, the wings swing back into a delta shape.

Speed versus range, payload traded for speed or distance—such are the constraints that confront every airplane designer. They are resolved by judiciously modifying the shape and power of the airplane, and bringing the forces of lift, weight, thrust and drag into the desired balance.

In general, birds that fly rapidly have pointed wings to minimize drag. When pursuing prey, the falcon can sweep back its wings in a high-speed dive, attaining speeds of over 200 miles per hour. Like the falcon, the swing-wings of the supersonic F-14 Tomcat are extended for maximum lift at takeoff and landing, but are swept back to reduce drag at supersonic speeds. The computer-controlled, variable-sweep wings give the bulky F-14 surprising flexibility and mobility.

Long, narrow wings give the albatross a large sail surface in proportion to its weight, enabling it to glide for miles on a steady ocean wind. Likewise, the long, slender wings and lightweight, streamlined fuselage of the Voyager and modern sailplanes provide them with a generous amount of lift at little expense in drag.

Like a pheasant taking to the air with an upward leap, an RAF Harrier jet literally jumps from the ground. Broad, rounded wings, for maximum lift at low speeds, characterize vertical-takeoff-and-landing (VTOL) aircraft such as the Harrier. The pilot can change the direction of engine thrust by rotating the exhaust nozzles down to make the plane rise straight up like a helicopter, or by turning them to the rear for conventional flight.

101

From Blueprint to B-747

A worker installs floor panels inside the shell of the fuselage. The spars and ribs of the fuselage are visible overhead.

From a 90-foot perch, a crane operator lowers the completed rear section of the fuselage onto an adjustable cradle. The cradle supports the weight of the section and enables workers to align it precisely with the center section of the plane.

Jumbo jet—the word is synonymous with Boeing's 747 line of commercial airliners. More than 700 of them are in service around the world today, and by the year 2000, at least three billion people will have flown on them. The newest model—the 747-400—incorporates the latest advances in the fields of digital avionics, aerodynamics, and structural design—all intended to increase the plane's range while decreasing its fuel consumption and operating costs.

Assembly takes place at a sprawling 1,000-acre plant in Everett, Washington. The plant itself is 2,000 feet wide and 1,600 feet long, and contains 291 million cubic feet of work area; it rises 11 stories high and covers the space of 47 American football fields. The power required to operate and maintain the building is enough to light more than 32,000 average homes.

The complex assembly sequence begins with the construction of the wings, which span 211 feet from tip to tip—almost twice the distance of the Wright Brothers' first flight at Kitty Hawk. The framework of each wing—its spars and ribs—is made of lithium-aluminum alloys and covered with sheets

Towering 63 feet above the floor, the height of the tail is equivalent to a six-story building. A fuel tank in the horizontal stabilizer adds an additional 350 nautical miles to the plane's range.

A technician guides an engine onto its strut. Including the cowling, it has a diameter of 8 1/2 feet. Each of the four high-bypass turbofan engines delivers up to 58,000 pounds of thrust.

A technician inspects a six-foot-high winglet—the most noticeable feature of the new B-747-400's wings. Jutting out at a 29-degree angle, the winglets control the wing tip vortices—the swirling mass of air that comes off the ends of the wings when they are producing lift.

More than 10,000 people work around the clock to complete up to five 747-400s per month, on two production lines. The work area is so vast that Boeing supplies a fleet of over 200 bicycles for employee transport and railway spurs come right into the building.

A technician inspects a shock strut of a wing landing gear just prior to installation. Each of the four main landing gears—two on the body and one on each of the wings—has four wheels. The impact of the 300-ton plane at landing is absorbed evenly by the shock struts, which act as pneumatic springs, and the 16 wheels, which are filled with nitrogen.

The pressure bulkhead rests on the floor prior to installation. This unit, together with a rear bulkhead, keeps the cabin pressure at a comfortable level at high altitudes.

Sitting behind an integrated glass cockpit, a two-person crew monitors computers that control the flight. Advanced digital technology has reduced the number of conventional instruments from 971 to 365. The airborne systems and flight status, displayed on six large colored cathode ray tubes (CRTs), reduce the pilot's workload and make him more of a systems manager than a stick-and-rudder pilot.

Fuel distribution system
Couplings, pumps and crossfeed valves interconnect all the fuel tanks. The 747-400 gulps six gallons of aviation fuel per mile, getting 0.16 mile to the gallon.

Wing tank
Two main tanks and a reserve tank in each wing store 36,700 gallons of fuel.

Main tank
Holds 17,164 gallons of fuel and is the core of the fuel system.

Air conditioning
Circulates up to 10,000 cubic feet of air per minute. When the plane is fully pressurized, approximately a ton of air is added to its weight.

Passenger emergency door
One on each side of the upper deck and five on each side of the main deck.

Upper deck
Seats 52 business-class or 69 economy-class passengers.

Flight deck
Seats the pilot and copilot and two observers; an air crew rest bunk is located behind the deck.

BOEING 747

Radome
Houses the weather radar, localizer and glide slope aerials.

Nose gear
Hydraulically actuated, the nose gear retracts forward into an undercarriage bay.

Forward hold
With a capacity of 2,800 cubic feet, can hold 50,000 pounds of luggage.

Water tanks
Located in the rear of the forward hold compartment, the tanks store 321 gallons of potable water.

Rear freight hold
Holds up to 2,340 cubic feet of containerized freight cargo.

THE FINISHED PRODUCT

The newest member of the 747 fleet, the 747-400, is ready to join the work force of commercial aviation. Weighing about 850,000 pounds, the plane can carry 250,000 pounds of fuel, giving it a total nonstop range of more than 7,000 nautical miles—about one-third of the Earth's circumference.

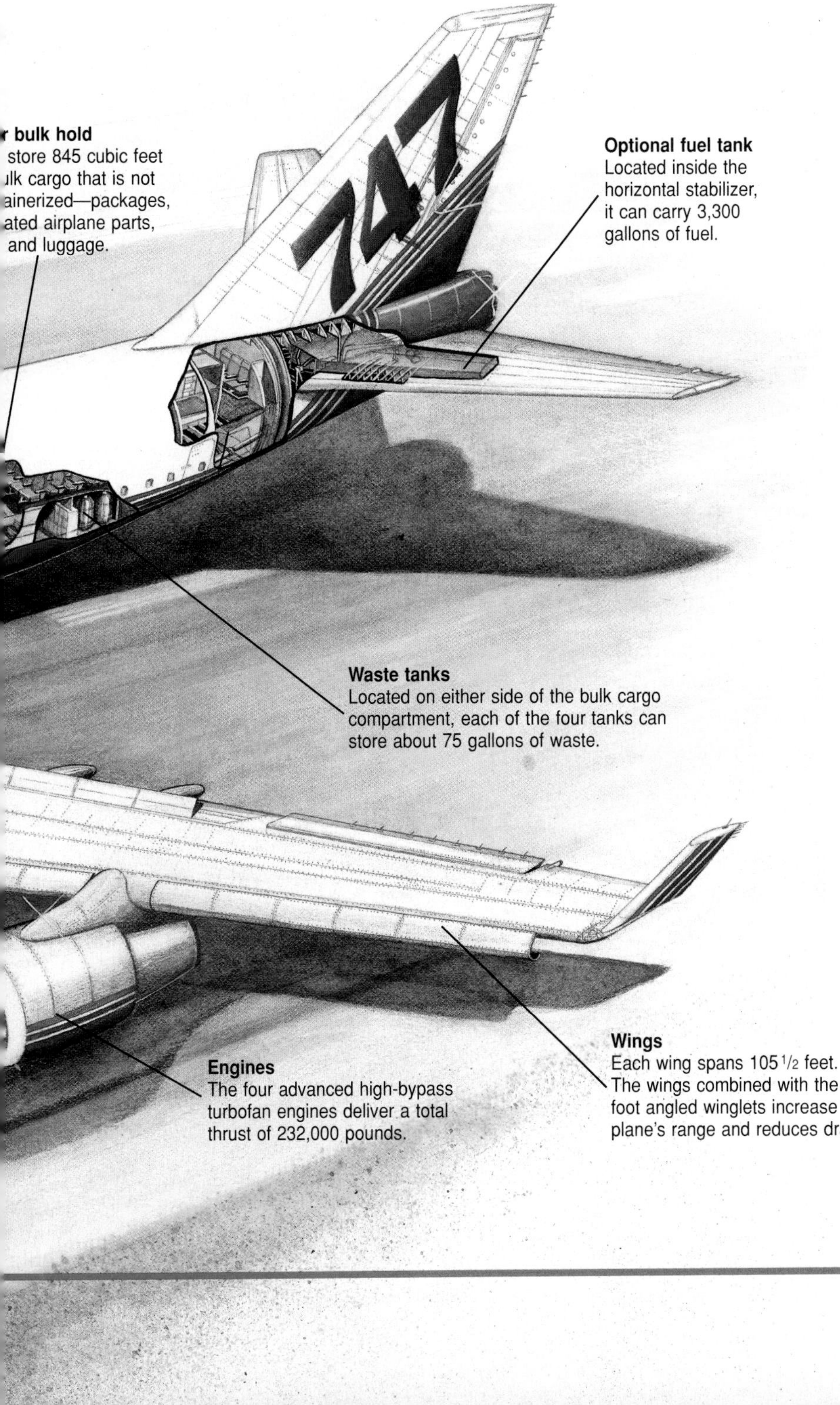

r bulk hold
store 845 cubic feet
ılk cargo that is not
ainerized—packages,
ated airplane parts,
and luggage.

Optional fuel tank
Located inside the horizontal stabilizer, it can carry 3,300 gallons of fuel.

Waste tanks
Located on either side of the bulk cargo compartment, each of the four tanks can store about 75 gallons of waste.

Engines
The four advanced high-bypass turbofan engines deliver a total thrust of 232,000 pounds.

Wings
Each wing spans 105 1/2 feet. The wings combined with the six-foot angled winglets increase the plane's range and reduces drag.

of honeycomb panels. Since the wings carry the weight of the plane, they are joined to a central wing unit and the plane is built around them.

The fuselage is assembled in three main sections. Its ribs, mainframes, and the one-half-inch outer skin are constructed of lightweight aluminum alloys. The center body section is lowered over the wing assembly by cranes and riveted to it. Next, the front and rear sections, assembled in other areas of the plant, are delivered to the final assembly area by an overhead crane. They are lowered onto adjustable cradles for precise alignment with the center section, then attached to it. With the body now complete, the horizontal stabilizer, tail cone and fin are attached to the rear section.

More than 100 miles of electrical wiring for the air conditioning, video, intercom and other flight systems are run through the 225-foot-long fuselage. Next, layers of woven, fire-resistant materials that insulate the fuselage are placed between the outer skin before the interior panels are installed. Only 7 1/2 inches of wall separates the passengers from the outside world. Finally the interior features are added—the stow bins, curtains, seats, toilets and galleys.

From the main site, the fully assembled plane goes to a paint hangar, where up to 600 pounds of paint (three coats) are applied to the exterior. Prior to its first flight, personnel check and recheck all systems. It is then flown through a series of rigorous flight tests. The plane is not delivered; once testing is completed, airlines must send a flight crew to pick it up.

MANEUVERING

Each year in August a swarm of pilots, airplane designers, backyard builders and flight buffs descends upon the quiet, rural city of Oshkosh, on the shores of Wisconsin's Lake Winnebago. The event is an eight-day celebration of aviation skill and technology known as the Annual International Fly-In Convention of the Experimental Aircraft Association. High above Oshkosh's Wittman Field, teams of stunt fliers like the one at right perform dizzying sequences of loops, spins, stalls, dives, barrel rolls, figure eights and other maneuvers that seem to defy all laws of gravity and aerodynamic sense.

Loudspeakers carry a stunt pilot's voice to the crowds on the ground. The words come in clipped, staccato bursts, taut with concentration and danger: "Nose down . . . full power . . . I'm leveling off . . . pulling up . . . harder!" Fifty feet above the runway, a tiny Pitts biplane snaps upward into a tight vertical spiral. At the top, as it breaks out of the maneuver, the plane flips momentarily onto its back, then rolls into a shallow dive. The forces of acceleration tug at the pilot's body with perhaps six times the pull of gravity, draining the blood from his brain.

None of this effort is visible from the ground. The plane sweeps and dives through its aerial ballet with the easy grace of an eagle cruising for its dinner, and the effect is a gasp of sheer wonder from the crowd. How can an aircraft corkscrew upward in this manner, or loop and roll, or even fly upside down, and not spin out of the air?

Pilots have been astonishing spectators with this puzzle since aviation's very first decade. Learning to guide an airplane's path was a basic step toward powered flight—as the Wright Brothers proved at Kitty Hawk *(page 36)*. The Wrights were soon followed by America's most flamboyant aerial showman, the designer Glenn Curtiss. On July 4, 1908 he flew a trim little biplane—the *June Bug*—through the six basic maneuvers of flight: takeoff, climb, turn, level flight, descent and landing, to win the prize offered by *Scientific American* to the first pilot to fly a measured distance of one kilometer. The air age had begun in earnest.

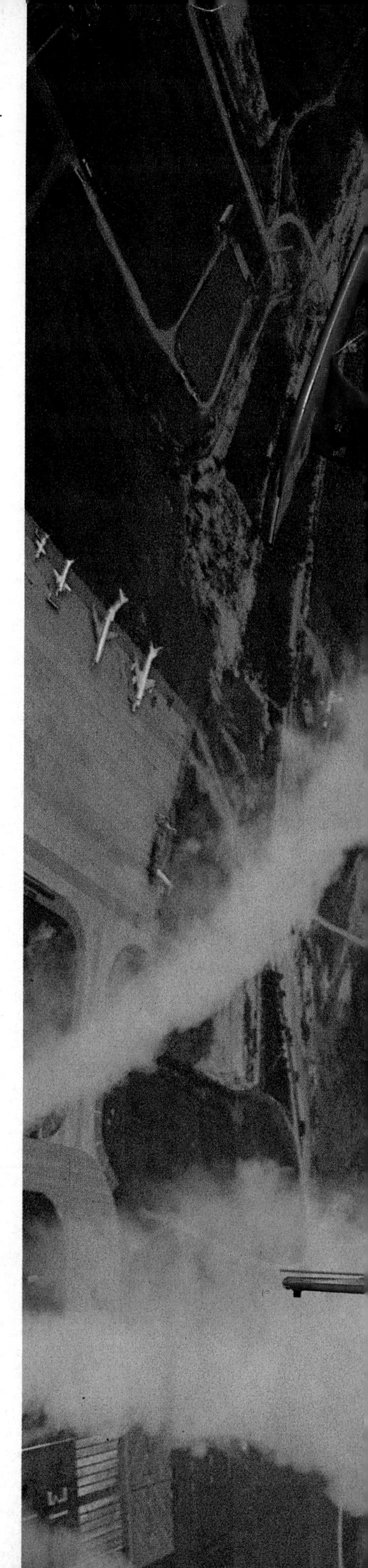

The Red Devils, an aerial stunt team, loop out of the clouds with gravity-defying grace. The team's three members, each flying his own home-built Pitts Special biplane, regularly perform at air shows across America.

Nature's Acrobat

No insect is a more accomplished aerialist than the common housefly. Performing loops, rolls, hairpin turns, flying backward, sideways, and vertically, it demonstrates flying techniques that involve extraordinary in-flight control.

During flight, the fly consumes prodigious amounts of oxygen, yet it has no lungs, nor does it breathe from the mouth or nostrils. Instead symmetrical rows of tiny air ducts run along its body. As the insect beats its wings, the muscles contract and force air out of the ducts; as the muscles relax, fresh air rushes in.

The power for flight comes from the insect's thorax, a box-like structure that houses the flight muscles. The wings are double-jointed, hinged to the thorax in such a way that they are free to move in any direction, the coupling system functioning like a ball-and-socket joint. Thus the insect rows through the air, beating its wings furiously in a figure-eight pattern like a sculling oar propelling a boat. To climb like a helicopter, it alters the angle of attack of its wings so that the current of air is directed downward rather than backward, just as the angle of attack of a helicopter's blades are altered in unison for vertical flight.

Unlike birds, which have a relatively weak upstroke, the fly produces equal propulsive power on the upward and downward strokes. To flap its wings about 200 times per second, the fly relies on a sophisticated mechanism in the thorax that boosts the speed of each wing beat. When the wings are midway through an upstroke or downstroke, the double-jointed wing couplings push against the elastic walls of the thorax and force it outward. As the wings continue through the stroke, the tension on the thorax is released. This forces the wings up or down with a snapping action, much like the action of a rubber band when it is being stretched and released.

Although most insects have two pairs of wings, flies and mosquitoes have given up the use of their hind wings. These have evolved into two small stumps, called halteres, that act like miniature gyroscopes. With the wings flapping, the halteres vibrate through their horizontal and vertical axes and send signals to sense organs at their bases. Any change in direction affects the normal undulations of the halteres and the fly is informed of its new flight course.

Its six legs are engineered so that the fly can land from almost any angle without slowing down. Instead it alights with a sharp jolt, each leg acting as a shock absorber. Unlike an airplane, the fly never runs forward after touchdown.

A multiflash photograph reveals how a fly lands upside down on a ceiling. Coming in for a landing, it approaches the ceiling at a 45-degree angle with its six legs fully extended. The two front legs touch down first, then it deftly cartwheels onto its other four legs to complete a perfect touchdown.

*This sequence of photographs displays a housefly (*Musca domestica*) at takeoff. As its middle pair of legs lift off the twig (*second photograph*), a starter muscle automatically triggers the main flight muscles and the fly's wings begin flapping. The wings rotate once through a figure-eight pattern and the back legs lift off.*

THE DYNAMICS OF CONTROL

The ability to make controlled maneuvers is essential to flight in any form—whether the fliers are birds, or stunt planes, or even humble insects *(page 66)*. There is no mystery to the process. In an airplane, a system of movable surfaces on the wings and tail allows the pilot to deflect the passing airflow, generating forces that pivot the plane in the desired direction.

From his seat in the cockpit, the pilot is within easy reach of all the necessary control devices. Each foot rests on a pedal, which connects to the rudder on the tail; his hand grips a control stick or wheel, allowing him to govern the elevators and ailerons. In most small airplanes the connection is mechanical—a basic "stick and string" arrangement as shown on the opposite page. Larger, high-speed craft require the extra muscle of hydraulics, which in some planes may even be activated by computers. But whatever the type of system, its operation is logical even to a non-pilot.

For example, a touch of pedal will angle the rudder to one side or the other—left pedal, left rudder, and vice versa. When the rudder swings over, it diverts the flow of air rushing past the tail fin. Two forces then come into play. One is a simple push, as the deflected airstream sets up an equal and opposite reaction; the tail is shoved to the side away from the deflection. At the same time, the rudder's new position changes the shape of the tail fin, making it a curve. The result is an airfoil, which in effect "lifts" the tail toward the low-pressure area on the outer side.

Together, these two forces move the tail; left rudder kicks the tail toward the right. This in turn pivots the airplane so that its nose swivels to the left. In much the same way, the helmsman on a boat controls his heading by using the rudder. If he angles the rudder to the left, the stern swings right, and the bow points left.

The same principle applies to the elevators on the tail's horizontal stabilizer. The pilot pulls back on the control stick—or, in larger planes, he pulls back the wheel—and the elevators tilt upward. This forces the tail down; the nose points up and the airplane climbs. By moving the stick from side to side the pilot manipulates the ailerons, one on each wing, which work in opposite directions. When the left wing dips the right one rises, and the plane begins to bank. It all seems like common sense.

Even so, aviation's pioneers took a while to arrive at it. The earliest known control system survives as drawings by Leonardo da Vinci, the great Italian painter and inventor of the 15th Century. Leonardo sketched a number of whimsical flying machines, including one in which a man rides a wooden framework strapped to a pair of bat-like wings. Since the pilot needs his hands to flap the wings, he must control his direction by other means. So Leonardo provided a harness, which the pilot wears on his head. A cable leads back from the harness to a dart-like tailpiece, which consists of crossed vertical and horizontal fins. By nodding or shaking his head, the pilot can wag the tail in any direction—presuming, of course, the craft can get airborne.

The Wrights fitted their *Flyer* with an elevator in front and a rudder behind, giving them both in-flight stability and basic control over altitude. But to bank the plane for turns, they used the wing-warping technique they had observed in turkey vultures: bending the wings' leading edges with wires attached to the wing struts. Ailerons would have accomplished the same job; indeed, a patent for them already

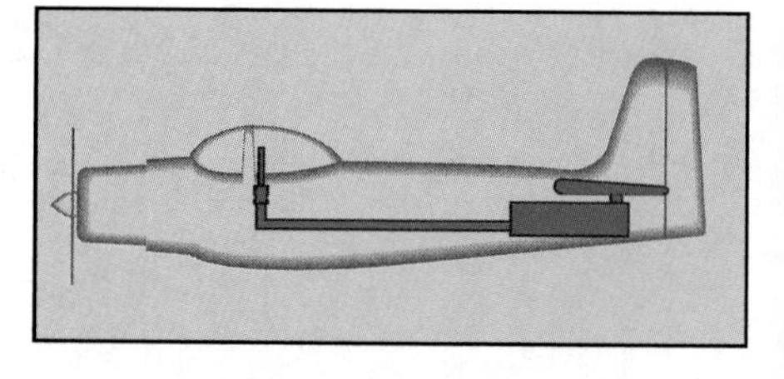

HYDRAULIC POWER

In a hydraulic system, manually operated cables lead to hydraulic activators, which do the work of moving the rudder, elevators and ailerons. A valve in each activator adjusts the pressure of hydraulic fluid, moving a piston connected to the appropriate control surface.

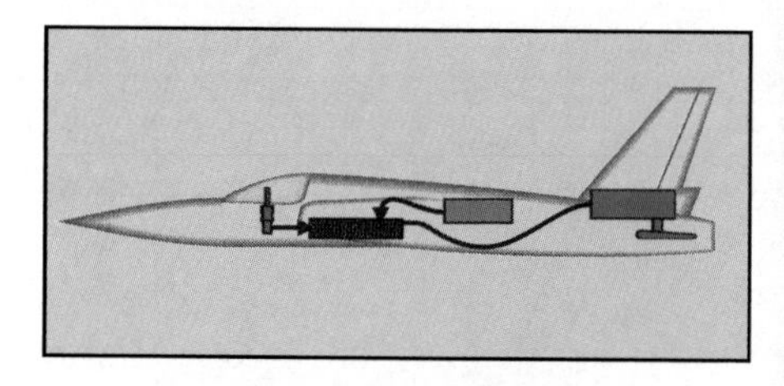

FLY-BY-WIRE

The most advanced control system uses electronic impulses to convey the pilot's commands. An on-board flight computer sends digital instructions to hydraulic activators, which in turn move the various control surfaces.

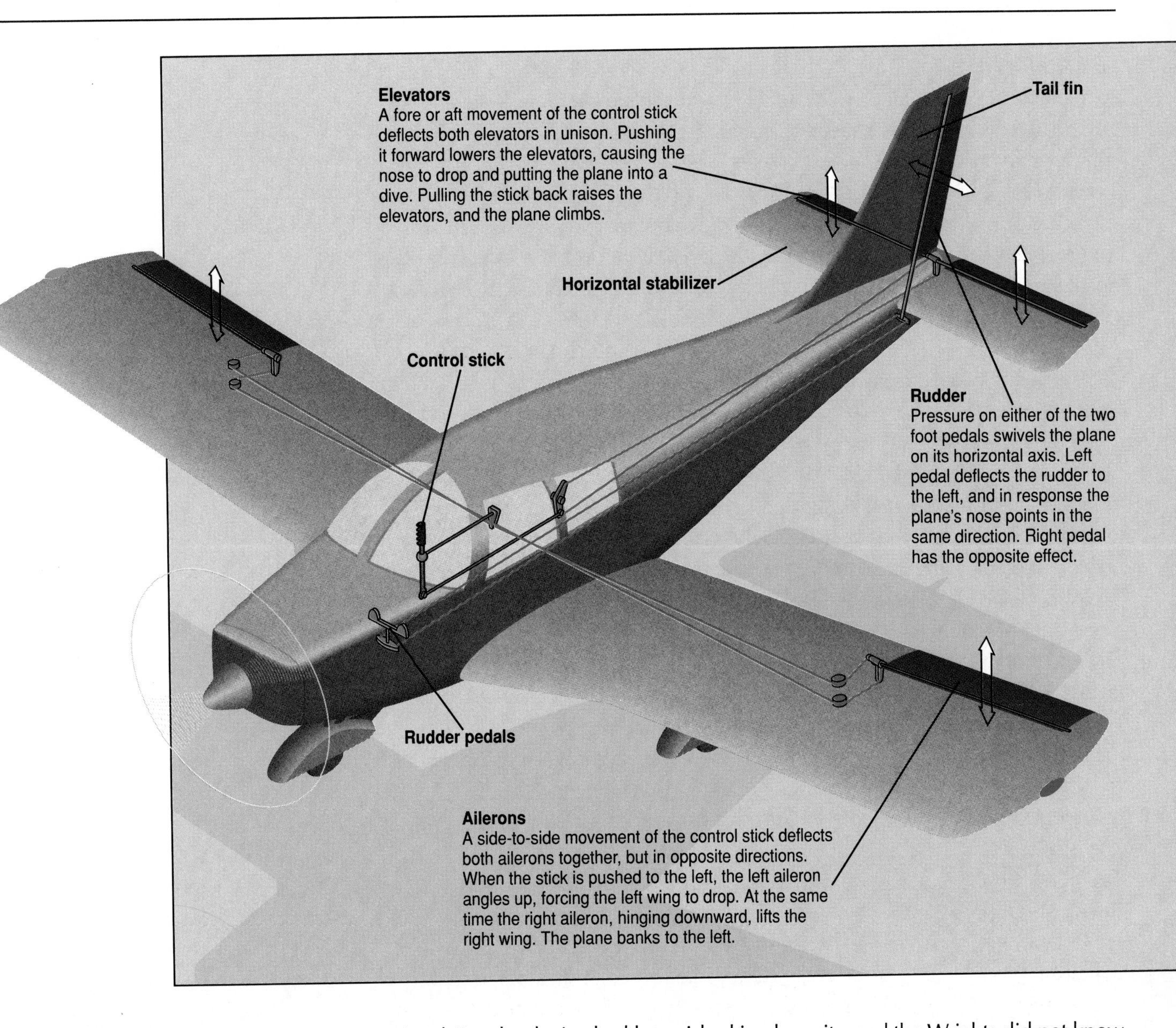

THE FLIGHT CONTROLS

The cockpit controls in this Piper Cherokee are linked to the external control surfaces through a system of cables, pulleys and levers: The rudder pedals connect with the rudder; the control stick with the elevators and ailerons. They allow a pilot to maneuver the plane in three directions—swinging its nose horizontally, pitching it up or down, or wagging its wings from side to side.

existed. But the device had languished in obscurity, and the Wrights did not know of it. What they did learn, by trial and error at Kitty Hawk, was something more basic: that neither the rudder alone, nor a dip of the wings, will cause an airplane to change direction. Only when the pilot uses both together can he put himself into a controllable turn. More than any other discovery, it was this that enabled the Wrights to fly.

Over the next few decades, the control of airplanes was considerably refined. By World War I most craft were equipped with ailerons. Then, in the 1920s, designers hit upon other devices to make the pilot's task easier. Trim tabs—miniature flaps on the trailing edges of the rudder and elevators—allowed pilots to adjust or "trim" their controls to a desired setting and leave them there. The tabs relieved pressure on the control surfaces, so that a pilot could remove his hands from the stick or wheel to perform other duties. The tabs also reduced the muscle power required to manipulate the controls.

The most recent advances in maneuvering are new systems for linking the cockpit controls with the external control surfaces. As airplane size and speed increase, so does the force of air against the flaps; eventually it becomes impossible to manipulate the flaps manually. Most planes today have hydraulic systems, much like an automobile's power steering. In some, the pilot's manual commands are amplified by means of hydraulic boosters, while in others hydraulics provide all the muscle. Should the main hydraulics fail, a backup system takes over. If that also goes, the pilot can still manually wrestle the plane into obedience. There is a price to pay, however; just as power steering dulls a driver's sense of the road, so hydraulic controls deprive the pilot of his "feel" for the air. To restore it, designers usually build in a system of springs or other simulators that convey a sense of the force pushing on the flaps.

The latest wrinkle is electronic. A small control column on the cockpit console—it resembles nothing so much as a computer-game joystick—sends electronic signals to one of several computers that control the appropriate flaps. The new Airbus A320 jetliner employs this "fly-by-wire" technique, as do such supersonic military craft as the Stealth bomber and the experimental X-29 jet fighter. The system has one main drawback. Strong electrical disturbances—lightning, for example, or even a powerful radar signal—can interfere with a plane's electronics. Eventually manufacturers may switch from wires to fiber-optic cables, which substitute pulses of light for electronic commands.

DEFYING GRAVITY

In flying any airplane, whether it be a commercial jetliner or the Piper Cherokee depicted in computer graphics on these pages, the pilot follows the same basic

TAKING OFF

With the plane's nose pointing into the wind, the pilot opens the throttle and the plane rolls forward, gathering speed. At the point where lift almost equals the plane's weight (about 65 miles per hour for a Piper Cherokee), the pilot raises the elevators. The tail drops, the nose tilts skyward, and the plane lifts off the tarmac. The steepness of its climb generates a heavy load of drag, slowing the plane down. To compensate, the pilot lowers the elevators. Leveling off, the plane builds up speed, allowing the pilot to raise the elevators once again for the climb to cruising altitude.

procedures, and uses his controls in virtually the same manner. Before taking off, he checks the weather, plots his route, and confirms his flight plan with the control tower. Once in the cockpit, he makes a thorough check of all systems. Are his flight instruments in order? Do his flight controls function as they should? Are his fuel tanks topped off?

On a large airliner like the Boeing 747, where the systems are more complex, the flight preparations are far more extensive. More than an hour is consumed by weather briefing, flight planning, assessing the fuel load and calculating such factors as weight and balance. Then, strapped in their seats, the captain and first officer systematically run through a checklist of almost a hundred items—electrics, hydraulics, pneumatics, pressurization, controls, instruments, radios, navigational equipment. "External power?" the first officer queries. The captain scans his power system voltmeter and advisory light, which show that the correct current is feeding the airliner's electrical system. "On and checked," he replies.

A dull moan announces the start of the first engine, quickly followed by the second. The elaborate checkout of the systems continues as the airliner taxis toward the runway for takeoff. "Starting systems?" . . . "Off." "Hydraulic panel and pumps?" . . . "Checked and on." "Generators?" . . . "Checked and parallel." As the plane approaches the takeoff point, the captain extends auxiliary flaps on the trailing edge of each wing, and a set of leading edge slats. With these moves, the airliner's streamlined, high-speed wings are reshaped into deeply curved, low-speed airfoils—essential for generating sufficient lift to overcome the jetliner's weight during takeoff.

Cleared for takeoff, the captain advances the four power levers; the engines roar and the plane begins to accelerate down the runway. As it picks up speed, the lift on its wings increases, and the plane feels eager to fly. When the airliner reaches 180 miles per hour, the captain eases back on the control wheel, angling the elevators upward into the airflow. In response, the plane's tail is forced slowly down and the nosewheel rises off the ground. This steepens the angle at which the wings attack the air. To overcome the increased curve, the air rushing over the wings speeds up still more, generating greater lift. Finally the upward force is sufficient to overcome the jetliner's weight, and the 390-ton plane rises majestically from the runway.

As soon as the plane is airborne, the captain momentarily levels off, with a resulting increase in forward momentum. In effect, he is giving himself the running start he needs to continue climbing. Watching his attitude indicator, which monitors the wings' angle of attack, he resumes his climb. He retracts the landing gear, reducing drag and gaining a further boost in acceleration. As the plane moves faster, its wing flaps and slats are no longer required, and they too are retracted. Pegging the airspeed at 270 miles per hour, the captain settles the plane into a smoothly rising climb and engages the autopilot.

ON THE STRAIGHT AND LEVEL

The airliner will make most of its journey at a cruising altitude of 31,000 feet, and the captain has already punched this information into his autopilot. About twenty minutes into the flight, as the plane nears the desired level, an amber light flashes. "Approaching Flight Level 310," the first officer announces. The next step is automatic. The autopilot lowers the nose of the plane into the cruising attitude and the airspeed increases. The airliner settles into a comfortable cruising speed of about 620 miles per hour.

CRUISING

Moving ahead at a constant altitude, a pilot must coordinate his speed with the angle of attack. To maintain sufficient lift at slow speeds, he raises the elevators, angling the wings upward into the airflow. When pushing forward at a faster clip—and thus generating more lift—he can level off to horizontal. At maximum speed, he must deflect his elevators and move in a slightly nose-down posture (opposite page)*; otherwise he will find himself gaining altitude. Cutting his throttle and again leveling off, the pilot finds his most efficient cruising attitude.*

Outside the plane, invisible aerodynamic forces are at work. During acceleration, when thrust was greater than drag, the throttles were set to generate full engine power. The elevators were tilted upward to keep the plane climbing. Now, the autopilot signals minute nose-down corrections, lowering the elevators and edging back the throttles. Seeing that his electronic assistant has accurately pinpointed height, speed and outbound track, the captain sets the autopilot to cruise—a maneuver called "letting George do it."

With the aircraft stabilized in straight and level flight, the four aerodynamic forces are in perfect balance. Thrust equals drag, and lift is just sufficient to compensate for the airplane's weight. Should any force change, there will be a corresponding alteration in the plane's flight path. Thus, if the captain throttles back, decreasing thrust, lift will also diminish and the plane will start to descend. To maintain altitude he can open his throttle again. Or he can raise the elevators—increasing lift by pitching the plane's nose slightly upward so that the wings meet the air at a higher angle of attack. With the plane in a new equilibrium, it will move forward on a level path, but at decreased speed.

In practice, the automatic pilot will take care of most such adjustments. The occasional burble of turbulence as the plane passes through flecks of high-level cloud thus requires no action on the flight deck. Also, the plane's inherent stability tends to correct for minor disturbances in pitch and roll.

A routine fuel check shows that the engines are gulping almost 100 pounds of fuel per minute. In an hour's time, the airliner's weight will have decreased by nearly three tons. Less weight gives the captain lift to spare, and he can use it to fly more efficiently. As fuel is consumed, he dips the plane's nose slightly, lowering the angle of attack. Wing drag diminishes, and this translates into a bonus in thrust. The captain can then either fly ahead at increased speed, or he can cut back on power to save fuel.

BANKING AND TURNING

Glancing at his en-route navigation chart, the captain orders a course change: "Turn left onto 330 degrees." The first officer responds by gently swinging the control wheel to the left, putting the airliner into a banking turn. The left-wing aileron flips up into the airflow, generating a downward force that dips the wing. At the same time, the right wing rises as its aileron moves down, enhancing the wing's airfoil shape and increasing its lift. With the plane now banked at a 30-degree angle to the left, the turn begins. In much the same way, on a smaller scale, a bicycle rider can change his path by leaning to the left or right.

The force that pulls the airliner into the turn is a portion of its total lift. With both wings tilted, some of their lifting power is directed sideways toward the center of the turn—in this case, to the left. Thus the plane assumes its curving flight path. But other forces come into play, which tend to disrupt the maneuver. With its weight unbalanced to the left, the plane tries to sideslip. At the same time, with less lift pulling upward, the plane begins to lose altitude. The captain compensates by increasing the power, and by nudging up the elevators.

Another force causes the nose to swing toward the outside of the turn. This tendency, known as yaw, results from the extra lift on the rising right wing. By the laws of aerodynamics, more lift means more drag; this slows the right wing, holding it back from the turn. To bring the nose into line, the captain calls upon the rudder, the control device for correcting yaw. A touch of left rudder points the nose to the left, into the turn.

Used by itself, the rudder has little effect in changing an airplane's course. The nose will pivot, to be sure, but the plane continues forward in a kind of aerial skid—much like an automobile spinning on an icy road. Only when a rudder

MAKING A TURN

To turn smoothly without sideslipping, the pilot moves the rudder and ailerons simultaneously, banking the plane and swinging its nose in the direction of the turn. In this right turn, he uses right rudder and moves his stick to the right, raising the right-wing aileron while lowering the left. To complete the maneuver, he moves rudder and ailerons in the opposite direction to level the wings.

movement is combined with a banking maneuver do the forces act in harmony to turn the plane. But the amount of rudder pressure must be just right. Too little, and the yaw persists. Too much, and the plane can plunge into a disastrous spiraling dive from which it is difficult to recover.

With all three controls working together, the airliner swings toward its new course. The forces acting upon it are in perfect equilibrium. The inward lift that causes the turn is exactly matched by an outwardly directed centrifugal force. Vertical lift is equal to the airplane's weight; thrust balances drag. The result is a turn so smooth that no one in the plane can feel it. A pendulum would hang perpendicular to the plane's tilted floor, and a cup of coffee on a passenger's seat tray would not spill a drop.

As with the jetliner, so the Piper Cherokee shown below requires the same judicious application of ailerons, elevators and rudder to execute a smoothly controlled turn. There is one important difference, however. The Cherokee, light in weight and built for maneuverability, can snap through its turn in a matter of seconds. But a Boeing 747 has the unwieldy momentum of an airborne supertanker. It requires more than a minute and a mile or so of airspace to swing through an arc of 180 degrees.

THE THREAT OF STALL

As the passenger jet approaches a radio beacon en route, Air Traffic Control advises of congestion ahead and instructs the captain to cut his airspeed. He pulls back on the power levers, and the needle on his airspeed indicator drops. But the reduction in airflow also results in a loss of lift, and to maintain altitude he must increase his angle of attack by slanting up the nose. The aerodynamic forces rearrange them-

selves: Lift still compensates for weight, but a greater portion of it comes from the steeper angle of attack. The plane moves forward at a slower pace.

Cruising along at this high-pitched attitude is called "mushing," and there is a limit to how far a pilot can push it. At constant airspeed, raising the angle of attack from four degrees to eight degrees, for example, will about double the lift force. But when the angle of attack reaches about 14 degrees above horizontal, the airflow over the wings becomes disturbed and lift is destroyed. Eddies of turbulence break up the flow, and the difference in pressure above and below the wing begins to dissipate. By about 18 degrees above horizontal, the destruction of lift is complete.

PULLING OUT OF A STALL

Climbing too steeply to sustain the smooth flow of air over his wings, the pilot stalls. Raising the elevators to bring up the nose would only make matters worse. Instead, the pilot takes advantage of the dive by lowering the elevator and pushing the throttle. Powering forward at the lowered angle of attack restores the forces of lift, and the pilot regains command.

The plane goes into a stall: It begins to drop, nose down, with alarming rapidity. Most stalls are to some extent self-correcting. As the nose pitches down, the angle of attack declines, speed picks up and lift is restored. The pilot can assist by pushing the throttle forward and lowering the elevators—both lift-enhancing maneuvers. But this assumes that the plane has plenty of altitude, allowing the pilot to regain control before he crashes to the ground. All in all, the best strategy is to avoid stall in the first place.

Most commercial aircraft are equipped with stall warners. These devices, which are connected to the angle of attack indicator, sound bells, flash lights or emit loud beeps to alert the pilot to an impending stall. Should the angle of attack increase, the control column would begin to shake back and forth. Then, if none of these warnings were sufficient, an automatic stick pusher would literally thrust the control column forward, effectively changing a nose-up command into a nose-down maneuver.

TACTICS FOR LANDING

Coming in for a landing, the captain of the jetliner and the pilot of the Cherokee have the same goal: to touch down gently at the slowest possible speed. All kinds of factors can complicate the maneuver—wind and weather conditions, traffic patterns at the airfield, runway length, to name a few. Ideally, the pilot will land directly into the wind so that the onrushing air provides an extra bonus of lift and maneuverability. A crosswind makes his task more difficult, since it tends to slew the plane to one side or the other, called "crabbing." Constant adjustments of rudder, ailerons and elevators are needed to keep the nose of the plane pointed straight down the runway.

The Cherokee, light and slow-flying, can begin its approach with a minimum of preparation; its pilot need only alert the airfield's control tower and await instructions. The airliner crew must get ready for landing well in advance. Some 30 minutes or so before reaching the airport, the captain begins a slow, steady descent from cruising altitude. He radios Air Traffic Control and, if there are no other planes ahead of him, he is given clearance to land. At this point, he is still ten miles away from touchdown.

Disengaging the autopilot, the captain adjusts his course for the final approach. About seven miles from the runway he throttles back his engines, reducing airspeed; the nose dips and the plane enters a shallow dive. As the runway nears, the captain cuts the speed of the airliner way back, to around 160 miles per hour. To make up for the resulting loss of lift, he deflects the elevators, setting his wings at a higher angle of attack. He also reshapes the wings, extending their leading-edge slats and the flaps on the trailing edge—reversing, in effect, the procedure followed for takeoff. About 2,000 feet above the runway, the first officer lowers the landing gear, imposing an extra drag that helps settle the airplane into a steep, mushing descent.

Closing in on the runway, the captain notes the touchdown point, which rushes toward him at nearly 200 feet per second. He tilts up the nose until the plane is on the verge of stalling, and the plane seems to hover just above the tarmac. The captain activates a set of airbrakes, called spoilers, which emerge from each wing and create turbulence that destroys the remaining lift. The airliner touches down as the main landing gear makes contact.

With the plane firmly on the ground, a sudden roar announces that the captain has reversed the engines and applied full throttle. The plane quickly sheds its remaining speed, with only a touch of the wheel brakes—gingerly applied to prevent overheating—needed to maintain control. The captain moves his hand to the steering control that guides the nosewheel, ready to begin the slow taxi from the runway to the debarkation gate.

LANDING

The plane descends toward the runway at a steep angle, its engines turning slowly. Near the ground, the pilot raises the elevators slightly and revs the engines; the increased angle of attack and extra thrust help maintain lift. Less than 20 feet from the ground, power is reduced and the pilot keeps the plane airborne by slight increases in its angle of attack. With the wings almost at a stall, the plane seems to float momentarily above the ground. Then it softly touches down on the main wheels. The pilot quickly trims the elevators, then closes the throttle to cut speed even further. The nosewheel touches down, and a combination of reverse engine thrust and wheel brakes brings the plane to a standstill.

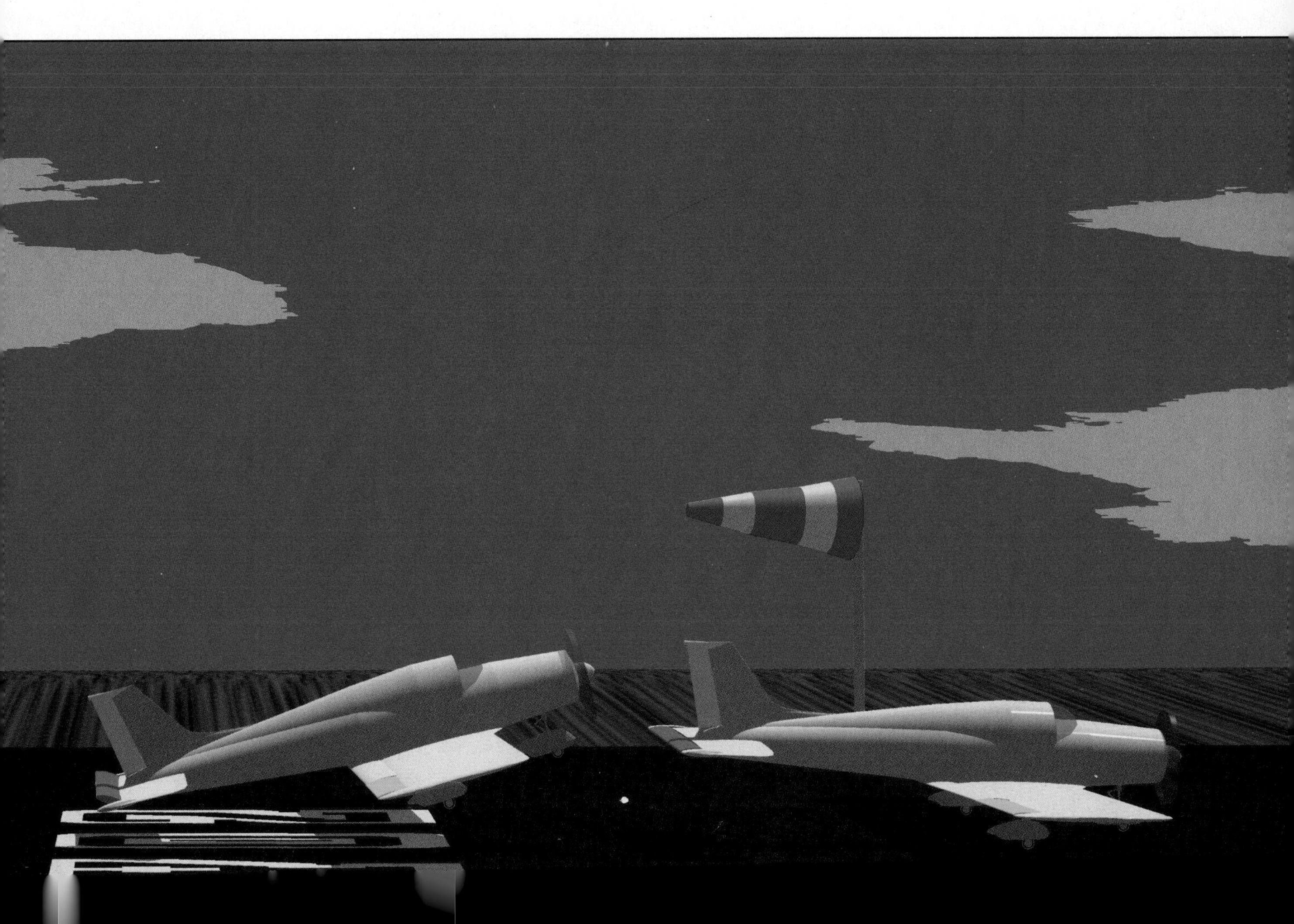

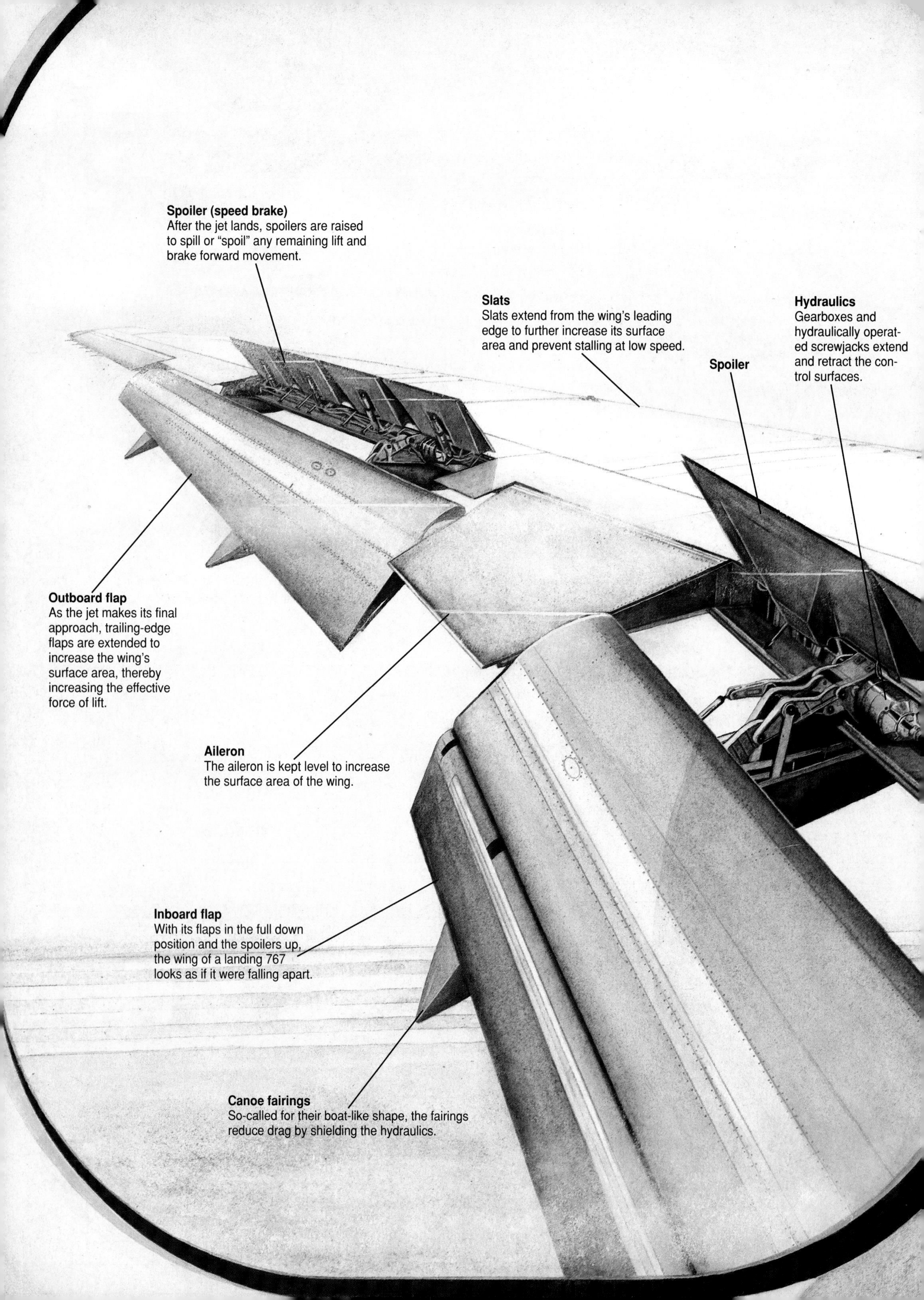

Spoiler (speed brake)
After the jet lands, spoilers are raised to spill or "spoil" any remaining lift and brake forward movement.

Slats
Slats extend from the wing's leading edge to further increase its surface area and prevent stalling at low speed.

Spoiler

Hydraulics
Gearboxes and hydraulically operated screwjacks extend and retract the control surfaces.

Outboard flap
As the jet makes its final approach, trailing-edge flaps are extended to increase the wing's surface area, thereby increasing the effective force of lift.

Aileron
The aileron is kept level to increase the surface area of the wing.

Inboard flap
With its flaps in the full down position and the spoilers up, the wing of a landing 767 looks as if it were falling apart.

Canoe fairings
So-called for their boat-like shape, the fairings reduce drag by shielding the hydraulics.

THE NAKED WING

Viewed through a passenger window at the moment of touchdown, the wing of this Boeing 767 is transformed into a slow-speed, high-lift device by means of flaps, slats and spoilers.

THE WING AT TOUCHDOWN

The most difficult period of any flight comes in the last few minutes before the wheels touch down. With the plane's speed slowed to a minimum, the pilot calls upon all his skill and concentration to maintain lift and avoid a disastrous stall. To help prevent it, airline designers have taken a cue from nature: Like a dove or an eagle, the 747 approaches the ground with its wings transformed.

When an eagle lands, it cocks its wings into the airflow and spreads its feathers, thus giving itself an extra degree of lift and control. The pilot of the 747, in extending the slats and flaps at the wings' leading and trailing edges, accomplishes the same goal. The wings are broadened so that they expose a larger undersurface to the air. At the same time they take on a more pronounced camber, which deflects the air sweeping over their upper surfaces into a speedier, more highly curving flow. As a result, air pressure increases below the wings, and decreases above them, for the needed boost in lift.

To a passenger with a window seat, the process may seem downright alarming. A wing before touchdown appears to self-destruct. The trailing edge flaps, mounted on guide tracks, slide downward and back in two, or often three, interconnected sections, activated by large hydraulic screwjacks. Spaces open up between each section, revealing gaps of naked sky. But this apparent disintegration is simply the mechanical equivalent of the eagle spreading its feathers. The openings actually augment the forces of lift by diverting additional airflow over the flaps' upper surfaces. Furthermore, the wing's total area increases by as much as 25 percent, virtually doubling the lifting power.

The final transformation occurs just before the airplane settles onto the tarmac. The spoilers, located directly ahead of the flaps, hinge upward like box lids and

What an airplane accomplishes with the help of mechanical devices, the bird does with natural grace and ease. Here an eagle inclines its body to catch the uplifting rush of oncoming air—thereby increasing its angle of attack—and extends its tail downward in a wide fan as an airbrake. At the same time, it spreads and twists its wingtip feathers to direct the airflow more effectively over the wings' upper surfaces. In the last few feet of flight, the eagle fans out its feathers to create a larger wing area, allowing it to float, parachute-like, to its perch.

push directly against the oncoming airstream. The sudden increase in air resistance kills the plane's speed and destroys its remaining lift. The spoilers thus serve the same function as the eagle's broad tail, which fans downward on landing to brake the momentum of descent.

MANEUVERING ON THE GROUND

When Louis Blériot made his pioneering flight across the English Channel in 1909, the mechanics of landing his craft were the least of his worries. His tiny monoplane weighed only 660 pounds, and he brought it back to earth on landing gear constructed out of bicycle forks, wire-spoke wheels and heavy rubber tires. Today's jumbo jets, touching down at speeds of up to 200 miles per hour, require far stur-

A telephoto freezes a Boeing 747-400 at the instant of touchdown. Descending at 170 miles per hour with its massive undercarriage lowered, the plane's 18 wheels slam into the runway with express-train force. The massive hydraulic shock absorbers take most of the impact. The 49-inch tires, inflated to 300 pounds per square inch, cushion the rest of the impact, and the airliner's carbon-disc brakes bring it to a gentle halt.

dier support. Their landing gear needs to be ultra-strong, relatively light in weight, and compact enough to retract neatly out of the way into the craft's slender wings during flight.

The main landing gear of a Boeing 747 is a near miracle of design ingenuity. It consists of four separate wheel trucks: one at the base of each wing, and two that retract into the plane's belly. Each truck carries four wheels rimmed with high-pressure, nonskid tires. Together they must support virtually double the 747's certified maximum takeoff weight of almost 900,000 pounds; as a fail-safe requirement, Boeing dictates that the jetliner must be able to land using only two of its main trucks. Two additional wheels, mounted on a truck extending from the nose, cradle the plane when it comes to a stop.

Toying with Aerodynamics

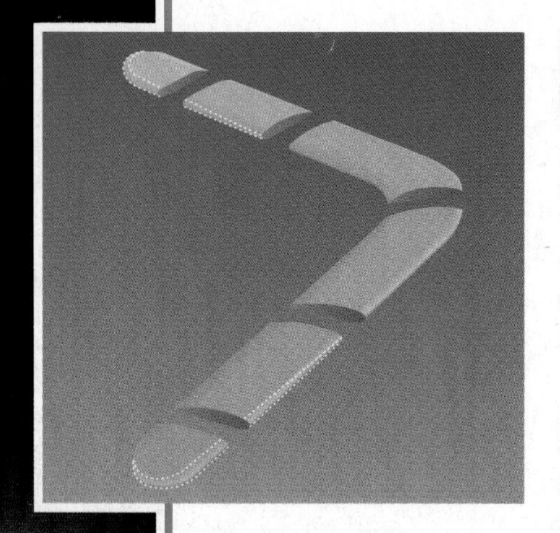

This cross section reveals the contoured top and flat bottom of the boomerang's two blades.

Captured by time-lapse photography, boomerangs circle the midnight sky, then float back to the ground like twirling maple seeds.

Humans have been tossing boomerangs since prehistoric times. The oldest example, a boomerang excavated from a cave in Poland, dates from 20,000 B.C. Neither is the flying disk a new invention. Natives of 18th-Century India threw metal rings as weapons. These aerodynamically complex devices have been transformed into intriguing toys by contemporary imaginations.

Scientists have used slide rules and computers in an effort to explain the flight of the boomerang. It works by a combination of lift, gyroscopic forces and Newton's Laws of Motion. When a boomerang is launched, a combination of linear and rotational velocities gives it forward motion, stability, aerodynamic lift (as well as drag), and gyroscopic precession *(page 100)*. Usually fashioned in the traditional V shape, boomerangs can take the shape of every letter of the alphabet except the letter "I." Once airborne, the boomerang becomes a set of rotating wings; the airfoil shape of the two arms *(inset)* creates lift like that of a helicopter's blades.

Gyroscopic precessional forces cause the boomerang to circle. In flight, the upper blade rotates into the wind, while the bottom blade rotates away from the wind, so that the top blade generates much more lift. This force causes a gradual rolling torque to the left (in the case of a boomerang thrown with the right hand). But instead of the boomerang actually rolling to the left, gyroscopic precession converts that torque into a left turn, like a "no hands" turn on a bicycle. Hence the circular flight path.

Unequal forces, created by the shape and twist of the blades, shift the center of lift slightly forward of the center of rotation. This results in a different kind of precessional force that causes the boomerang to gradually tilt over or "lay down" in a horizontal position midway through its flight. Finally the boomerang hovers, while still spinning, and drops into the hands of the thrower.

Recently, a mechanical engineer from California designed a flying ring with the help of NASA aerodynamicists and computers. Called the *Aerobie*, it has been hurled about a fifth of a mile—farther than any other man-made object without benefit of a power supply or external force like the wind. But although a flying ring may be thrown farther than a boomerang, it has to be retrieved.

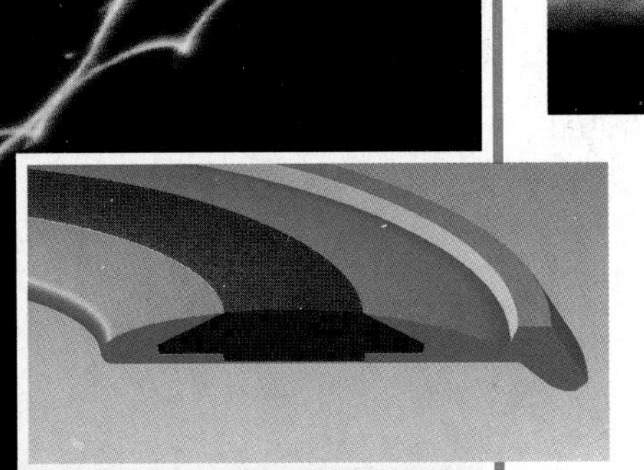

THE SECRET IS IN THE SLICE

The flying ring, weighing just four ounces and only a tenth of an inch thick, is made of soft rubber partially surrounded by a flexible plastic core (left). *It works like two airfoils in tandem. The leading half of the ring creates more lift than the trailing half, which flies in disturbed air. A ridge, or spoiler, on the outer rim brings the center of lift back to the center of rotation, gives the ring better balance, and reduces drag for straighter, longer flights. A spinning disk, although inherently stable, will always roll to the side as it moves through the air because its center of lift is usually forward of its center of rotation. This causes the disk to pitch up; the rotation converts this tilt to a sideways roll—either to the left or right.*

Most of the impact on landing is dissipated by heavy-duty shock absorbers attached to each wheel truck. In earlier aircraft, the shocks were generally heavy steel springs, but in the 747 and most other modern aircraft they consist of hydraulic cylinders filled with oil and either air or nitrogen. The trucks are hinged to powerful, hydraulically activated steel support beams that pivot down from their wing or fuselage housings on command from the cockpit. On the ground, the pilot rides three stories above the runway. Said one 747 captain, "It's like sitting at my attic window and trying to drive my house down the road."

Besides the forces of impact, each landing gear must also absorb deceleration loads of up to 64,000 pounds that occur when the pilot applies his wheel brakes. Assisted by reverse thrust, a 747 can screech to a stop in as little as 2,500 feet—a performance that absorbs as much kinetic energy as a million automobile brakes. Hydraulics again come into play, pushing multiple brake discs against each wheel drum. So intense is the friction that in some models the wheel assembly includes cooling fans to disperse the heat; otherwise the discs would soon glow a bright cherry red and cause the tires to burst. The plane's latest version, the 747-400, solves the heat problem with plastics: carbon-fiber discs that run cooler, last longer and save a valuable 1,800 pounds total over the earlier steel brakes.

Like all heavy aircraft, the 747 also incorporates an antiskid device in each wheel. An automatic brake-release mechanism balances the brake pressure against the wheel's speed of rotation, ready to ease pressure the moment a wheel threatens to lock. The pilot can stamp down hard on the brake pedal without fear of swerving off the runway, even when the surface is icy. Even the tire treads are specially designed for abrupt stops in less than ideal conditions. Since puddles of rainwater on the runway tend to make the wheels skate across the surface without gripping the pavement, the tires are engineered to forcibly pump the water aside. Some tires also have side ridges that act as spray deflectors, which keep heavy splashes from hitting the flaps or other vulnerable control surfaces.

HUMMINGBIRDS AND HELICOPTERS

No airplane will ever match the maneuverability of aviation's mechanical magic carpet—the helicopter. Commenting on its versatile performance, the inventor of the first practical model, Igor Sikorsky, said: "I didn't anticipate all the many uses helicopters find today. But I was sure an aircraft that could fly like a hummingbird would be immensely useful."

Indeed it is. For like a hummingbird, and also most insects *(page 66)*, the chopper can take off from a standstill, fly forward or backward, move straight up or down, dance to either side, or hover in midair. And the comparison does not end there. Both bird and whirlybird owe their aerial agility to a capacity for precisely angling the pitch of their airfoils—the wings of the hummingbird, and the rotor blades on the helicopter.

In a sense, the hummingbird is an evolutionary compromise between more conventional avians and the insects, whose nectar-sipping habits it shares. Whereas other birds flex their wings freely at shoulder, elbow and wrist, the hummingbird is equipped with long feathered paddles that are almost all hand. In function, they resemble the rigid fans of a butterfly or a honeybee. Virtually all movement comes from the hummingbird's shoulder, a joint of remarkable flexibility that allows the

WHIRLING WINGS

A hummingbird hovers motionless in midair, its body slanted steeply upward so that the arc of its wingbeat is almost horizontal. As the wings sweep forward on the downstroke, and upward on the backstroke, they bear a close resemblance to helicopter blades—though with the wingtips tracing a horizontal figure eight. A steady blast of air is forced downward by both strokes, producing continuous lift.

wing to twist from rightside up to upside down through almost 180 degrees. Powered by flight muscles that account for up to 30 percent of the hummingbird's total weight, they flap at a rate of 50 to 70 times per second—many times faster than any robin or wren.

As they flip through the air, the hummingbird's wings generate lift during both the upward and downward beats; no other bird can do this. During hovering flight, the wings move in a figure-eight pattern, eggbeater fashion, each one sweeping horizontally through one-third of a circle. On the downstroke they are positioned normally, their upper surfaces angled skyward and their leading edges cutting the air ahead. But on recovery they swivel, moving up and back with their leading edges facing to the rear. The whir of these oscillating motions resembles nothing so much as the spin of a helicopter's rotor—which also produces lift throughout its entire revolution.

To shift direction—which it can do in the blink of an eye—the hummingbird simply alters the plane in which its wings operate. To back off from a flower, it plies both wings behind its body. Darting forward, it beats its wings more vertically, giving itself a forward lift that can propel it ahead at up to 50 miles per hour. Additional control comes from the tail, which the hummingbird can spread, close, cup or angle in the propwash of its whirring wings. There is literally no direction in which it cannot maneuver. Banking into a sudden turn, it can even fly upside down for a brief moment.

The only creatures that rival the hummingbird in aerobatic prowess are the insects, which learned all the tricks some 100 million years earlier. Their wings vibrate at even higher frequencies than do the hummingbirds'—for bees, 200 to 300 beats per second on average, and an astonishing 1,000 beats per second for certain midges. The earliest known flying insect is the dragonfly, which dates from a swampy, sultry era of time some 345 million years ago known as the Carboniferous period. Some dragonfly species grew to enormous size, with wingspans of three feet. None of these giants survive, but their modern-day relatives can hover and dart just as ably as any hummingbird—or any helicopter. Indeed, with its long tapering body and its four slender, rotor-like wings, the dragonfly could have been the model for one of Sikorsky's first military choppers, the R-5, and its civil counterpart, the S-51. When the S-51 was manufactured under license in England, in 1946, the Royal Air Force dubbed it—what else?—the Dragonfly.

All appearances to the contrary, a helicopter is not simply an aircraft with a large propeller on its roof. Its rotor is in fact a set of wings that, powered by a piston or turbine engine, provide lift by spinning. Each blade is an airfoil, which connects to a central drive shaft through an elaborate hinge mechanism. Other mechanical devices, known as swashplates and control rods, allow the pilot to adjust the pitch of the blades, and also to tilt the entire rotor assembly in any direction. Like the hummingbird's swiveling shoulder joint, these mechanisms, which together comprise the helicopter's rotor head, provide the key to the helicopter's omni-directional maneuverability.

Three cockpit controls replace the standard stick and rudder pedals of a conventional airplane. A floor-mounted collective pitch lever near the pilot's left hand varies the pitch of the blades. With an increase in pitch each blade swivels so that its leading edge cuts the air at a higher angle of attack, for an overall increase in lift; the helicopter rises. For an extra upward boost, the pilot advances his throttle, which is incorporated directly into the collective pitch lever. Because helicopter flight involves the constant use of both hands, the throttle is a twist-grip on the lever's handle, similar to that on a motorcycle.

The second control arm is the cyclic pitch control, which is located in the familiar stick position. This governs the helicopter's horizontal movements. Should the pilot wish to move sideways to the left, he pushes the control stick in that direction. The command is transmitted to the rotor head, which tilts the whirring blades, like an enormous spinning pie plate, slightly to the left. The lift they generate becomes divided, as on a banking airplane. The upward force keeps the helicopter airborne, while the sideways force pulls it smartly to the left.

Other aerodynamic forces complicate the business of helicopter maneuvering. The strongest is a tendency of the spinning rotor to swivel the craft's fuselage in the direction opposite to the spin. This powerful torque, which results from the principle that any action calls forth an equal reaction, was a severe problem for early helicopter designers, since it made their craft virtually unflyable. They addressed it a number of ingenious ways. Some designs included two rotors, one mounted above the other and spinning in opposite directions. One particular model had a pair of short, fixed wings with the rotors rising from each tip. Large cargo-carrying helicopters still use paired rotors—one in front above the cockpit, the other spinning above the tail. But most passenger helicopters now carry a small

INSIDE A ROTOR HEAD

On a typical helicopter rotor head, the blades are attached to a central rotor shaft powered by the engine. Pitch control rods leading from the upper swashplate attach to the leading edge of each blade. A rotating scissors device turns the swashplate as the blades revolve. The lower swashplate, linked to the cockpit, can be raised, lowered or tilted. These movements, conveyed to the upper swashplate, control the pitch and angle of the blades.

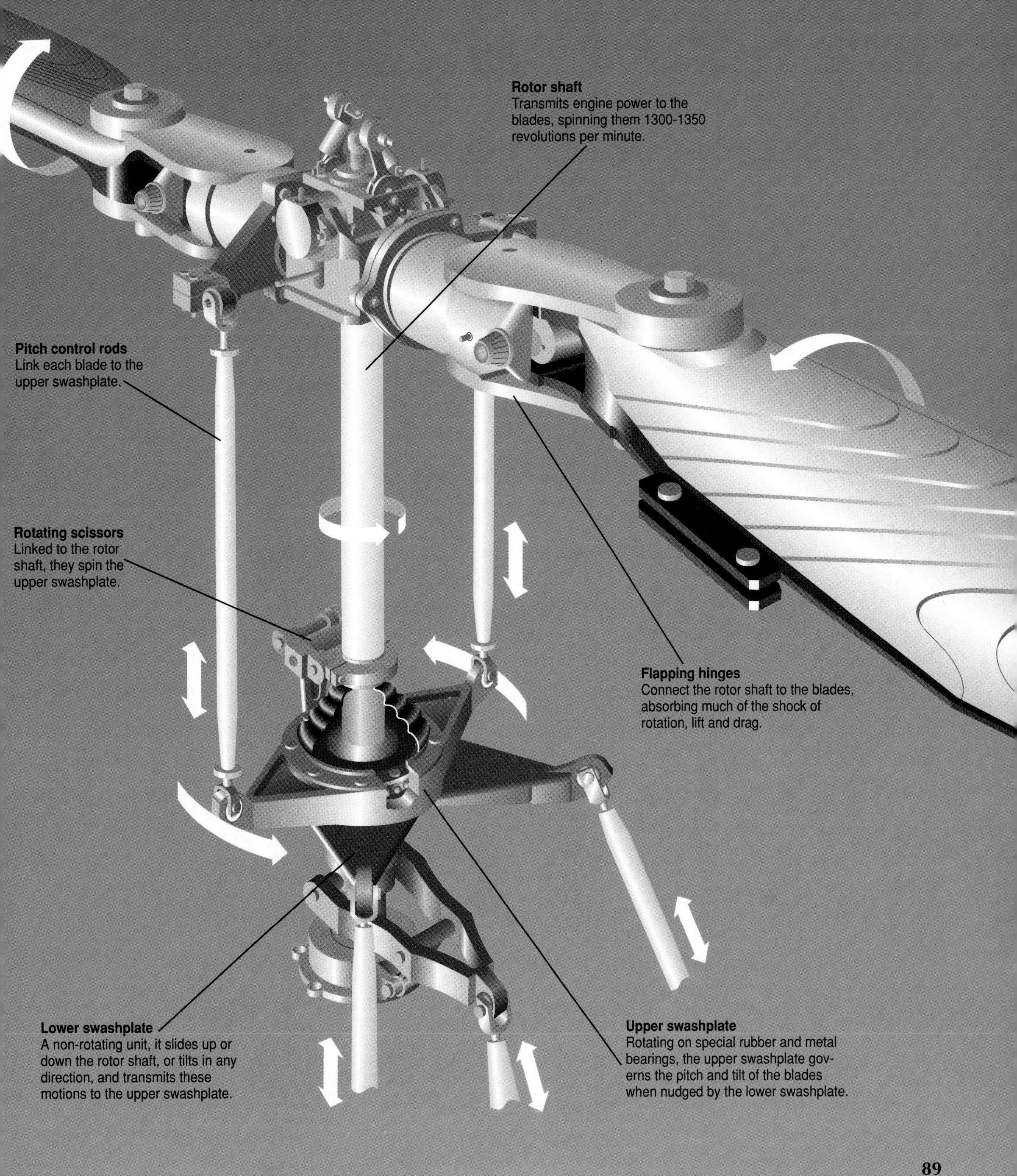

Rotor shaft
Transmits engine power to the blades, spinning them 1300-1350 revolutions per minute.

Pitch control rods
Link each blade to the upper swashplate.

Rotating scissors
Linked to the rotor shaft, they spin the upper swashplate.

Flapping hinges
Connect the rotor shaft to the blades, absorbing much of the shock of rotation, lift and drag.

Lower swashplate
A non-rotating unit, it slides up or down the rotor shaft, or tilts in any direction, and transmits these motions to the upper swashplate.

Upper swashplate
Rotating on special rubber and metal bearings, the upper swashplate governs the pitch and tilt of the blades when nudged by the lower swashplate.

tail rotor, mounted sideways like a propeller, which pushes the helicopter into line. It is governed by the third set of cockpit controls—the rudder pedals. These adjust the thrust of the tail rotor, allowing the pilot to steady his course, or steer to left or right during forward flight.

Another force takes effect whenever the pilot tilts the rotor assembly for horizontal movement. In the same way as a banking airplane develops more lift on its upper wing than on its lower one, the angled rotor blades are subjected to unequal pressures as they spin. In straight-ahead flight, with the assembly tilted forward, lift is diminished momentarily on the front-facing blade. Then, as the blade arcs backward, the lift upon it increases. It is this differential that pulls the helicopter forward; but it also puts a severe strain on the individual blade attachments. Thus the flapping hinges that connect the blades to the rotor shaft, which allow the blades to wiggle up and down, dissipating some of the force.

The pilot, one hand on the collective pitch lever, the other on the cyclic pitch control, his feet on the rudder pedals, can dart about the sky with almost total freedom, but the maneuvers take considerable skill. The hardest is hovering, in which he must pull just enough lift to balance the helicopter's weight, apply precisely the right amount of pedal to counter torque, and also hold the machine level by adjusting the rotor's tilt. It demands concentration, as one dyed-in-the-wool jet jockey found in his helicopter conversion course. "It's like trying to pat your head and rub your tummy while balancing on a bowling ball," he complained.

HOW A HELICOPTER FLIES

The pilot's cockpit controls are connected by a system of hinged rods to the helicopter's rotor head. The collective pitch lever sends commands to climb, hover or descend by moving the rotor swashplates up or down—thus changing the attack angle of the blades in unison. The cyclic pitch control, by tilting the swashplates, tilts the rotor head for movement forward, backward or sideways. Foot pedals govern the speed of the tail rotor; right pedal decreases thrust, left pedal increases it.

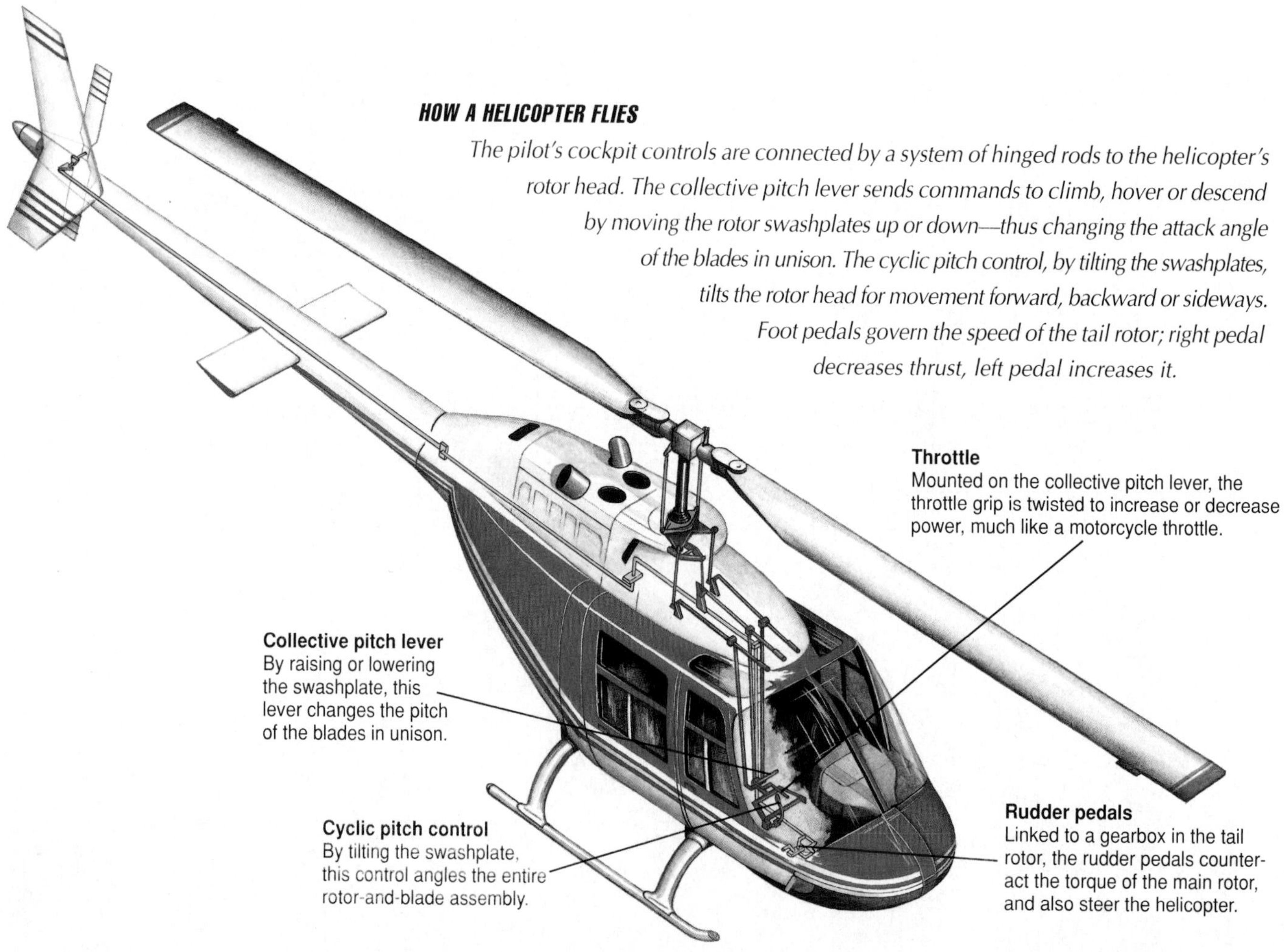

Throttle
Mounted on the collective pitch lever, the throttle grip is twisted to increase or decrease power, much like a motorcycle throttle.

Collective pitch lever
By raising or lowering the swashplate, this lever changes the pitch of the blades in unison.

Cyclic pitch control
By tilting the swashplate, this control angles the entire rotor-and-blade assembly.

Rudder pedals
Linked to a gearbox in the tail rotor, the rudder pedals counteract the torque of the main rotor, and also steer the helicopter.

Hovering
With the cyclic pitch control in neutral, the swashplate and rotor blades are level and the helicopter moves neither forward nor backward. The swashplate is raised to change the pitch, or angle of attack, of the blades in unison, producing lift to equal the weight of the helicopter and keep it motionless in the air.

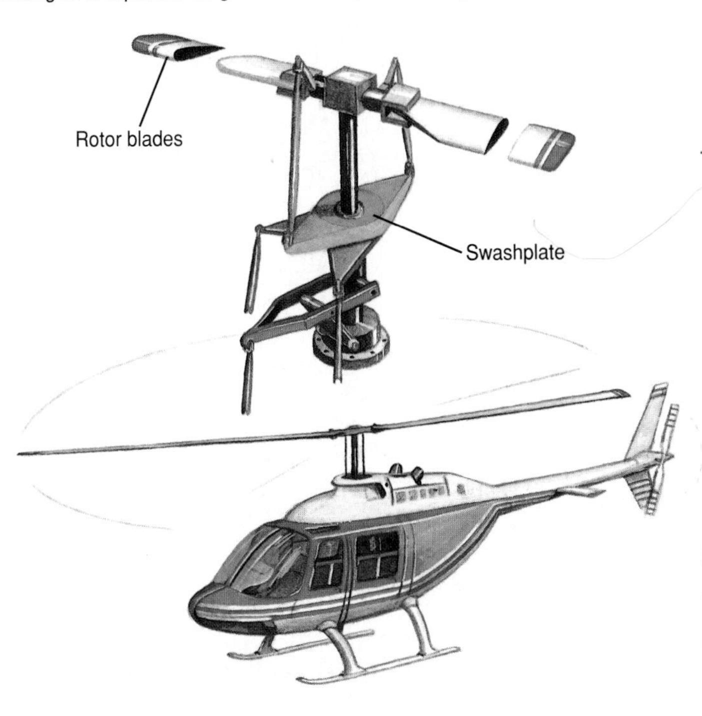

Forward flight
When the pilot pushes the cyclic pitch control forward, he causes the swashplate and blades to tilt down at the front. Lift becomes a divided force. It increases over the back of the rotor to support the helicopter's weight, and decreases at the front of the helicopter to provide forward thrust.

Vertical flight
To climb, the pilot pushes the collective pitch lever forward. This raises the swashplate and steepens the blades' angle of attack to the point where lift exceeds the weight of the helicopter. To descend, he reverses the procedure.

Backward flight
The pilot pulls the cyclic pitch control backward, tilting the swashplate and blades toward the back of the helicopter. With the force of lift divided—some directed up, some to the rear—the helicopter retreats.

The Amazing Dragonfly

Dragonflies, which have been around for more than 250 million years, may provide clues that could revolutionize aircraft design. Studies show that dragonflies use "unsteady aerodynamics," a mode of flying radically different from the flight of airplanes and birds.

Dragonfly wings churn up the air to create whirling vortices of low pressure that the insect uses to provide lift. By contrast, airplanes and birds rely on the steady flow of air over the upper and lower surfaces of their wings. For them, turbulence can be deadly. Taking a lesson from this remarkable flier, designers may someday imitate it by fitting aircraft with wings that would put unsteady airflows to good use.

The wings of an airplane, they suggest, could be fitted with devices that would redirect turbulence back onto the wings, decreasing pressure and increasing lift; this energy is now wasted. One idea is a flap or "fence" that would automatically flip up near the leading edge of a wing when the aircraft goes into an unexpected stall. This roughness would create vortices that increase lift, possibly averting a crash.

Another possibility is to install a small extra wing near the front of an airplane. This "canard" would mimic a dragonfly by sharply shifting its angle of attack to generate vortices as needed. The larger rear wing of the airplane could then coast along on these aerial whirlpools.

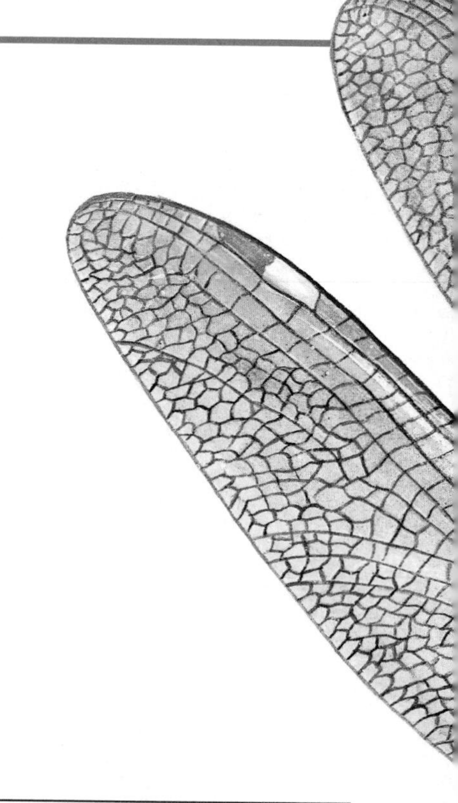

Four wings; four muscles
In the upper cutaway of a dragonfly's thorax a pair of direct flight muscles (dark) has contracted, resulting in the upward movement of the wings. In the lower diagram the second set of muscles has contracted; the wings flap downward. In this way each of the four wings can beat independently, making the insect highly maneuverable.

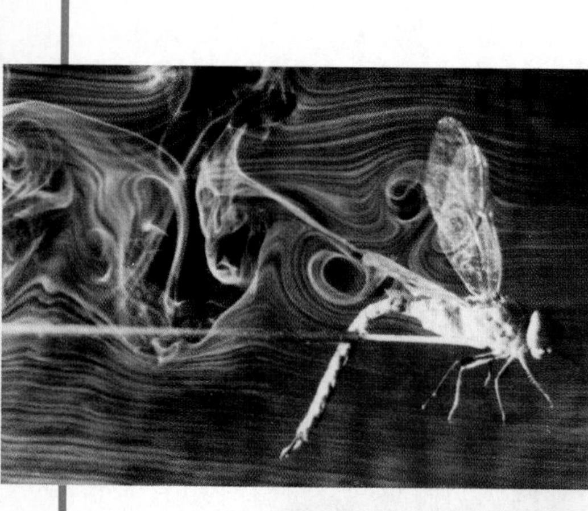

Articulated tail aids in steering and balance

With the help of miniature wind tunnels, stroboscopic photography and unwitting insect subjects, scientists are investigating the aerodynamics of turbulence. Here a dragonfly, tethered to a small wire, shows off its ability to generate lift by producing unsteady airflow. When the insect's front wings flap down, they stir up a small whirlwind which, when it passes over the back wings, generates enough lift to support up to 20 times the dragonfly's weight.

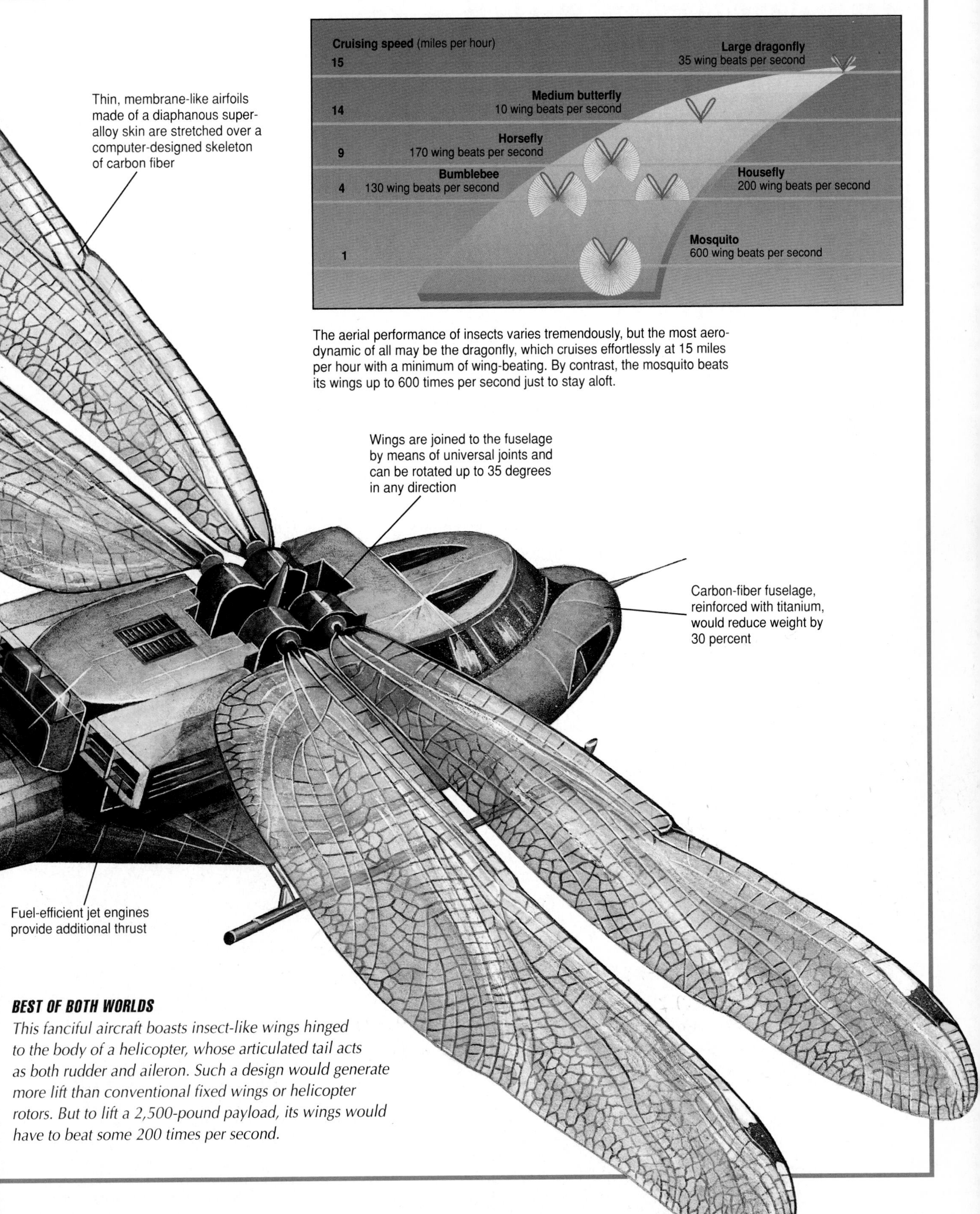

The aerial performance of insects varies tremendously, but the most aerodynamic of all may be the dragonfly, which cruises effortlessly at 15 miles per hour with a minimum of wing-beating. By contrast, the mosquito beats its wings up to 600 times per second just to stay aloft.

BEST OF BOTH WORLDS

This fanciful aircraft boasts insect-like wings hinged to the body of a helicopter, whose articulated tail acts as both rudder and aileron. Such a design would generate more lift than conventional fixed wings or helicopter rotors. But to lift a 2,500-pound payload, its wings would have to beat some 200 times per second.

THE CROWDED SKIES

No creature on Earth maneuvers in more hazardous circumstances than an ordinary bat. The only mammal capable of true flight, bats congregate in huge colonies, sleeping by day in caves or tree branches, then emerging at sundown by the tens of thousands to feed and frolic under cover of darkness. Some species leave their dens in close-packed swarms, as densely concentrated as raindrops in a thunderstorm, moving out along well-established flight corridors. Then they start foraging, in the blackest of moonless nights, some browsing on fruits, others flapping after insects so small as to be barely visible even in bright sunlight. The feeding is virtually continuous; a bat must consume from one-third up to half its body weight each night just to keep going. How does the bat do it?

The secret is a short-range navigation system as sophisticated as any in nature. Each bat emits a succession of shrill clicks, about 20 or 30 times every second, that beam out ahead and bounce back from any object they strike. The bat's sensitive, oversize ears pick up the echo. This aerial sonar allows the bat to zero in blind on the tiniest gnat or mosquito—and also to avoid colliding with trees, cave walls or other bats.

The clicks from a bat have such a high sonic frequency—up to 2,000 vibrations per second—that they are largely inaudible to human ears. But the energy level of each click is intense. At a distance of two inches from the bat's head, the acoustic pressure has been measured at 100 decibels, roughly the same as a jackhammer. As the bat flies, it alters the frequency and the length of each burst. Some bats broadcast simultaneously on two or more wavelengths. The more rapid the vibrations, the smaller the object the bat can detect. A mouse-eared bat, for example, can detect a steel wire only one millimeter thick from a distance of two yards. Then, as the bat moves in closer, its emissions become shorter and more rapid—with the brief intervals between each vibration allowing the echo to return. The time delay from a hungry bat to a mosquito six inches from its mouth is about one-thousandth of a second.

Borne on wings nearly as thin as a plastic bag, a multitude of bats emerges from a Texas cave, avoiding midair collision by means of aerial sonar.

The Migration of Monarchs

Warmed by the Sun, monarch butterflies fill the skies of central Mexico, searching for water and flower nectar. As the days lengthen, they begin to mate; then they depart, flying at speeds from 10 to 30 miles per hour on their northward journey.

Birds are not the only masters of long-distance migration: Fragile monarch butterflies cover more than 1,500 miles on their annual odyssey.

Since monarchs feed and lay their eggs only on the milkweed plant, one theory of their migration suggests that the butterflies originated in Mexico, where over half of the world's 100 species of milkweed are found. After the last Ice Age, the monarchs spread northward as the Earth warmed, following the milkweed. In returning to Mexico each winter, they are simply going home.

Scientists believe that the start of the migration is triggered by changes in the number of hours of daylight, the angle of light from the Sun and nighttime temperatures. Responding to these signals in the spring and fall, the monarch butterflies become restless and depart.

How do these insects find their way to the same summer and winter sites each year? Scientists theorize that, like some migrating birds, monarchs have a built-in compass that tells them their latitude and allows them to read the angle of inclination of the Earth's lines of magnetic force. En route, they feed voraciously on milkweed flowers, each butterfly adding about 100 milligrams of body fat to its delicate frame.

One of the largest wintering sites is a secluded enclave of the Sierra Madre mountains in Mexico. At an elevation of 9,000 feet, the average temperature hovers around the freezing mark. The monarchs, in a semidormant state, cling to the branches and trunks of 100-foot fir trees, the weight of their numbers literally bending the boughs.

In early March, the butterflies descend in swarms to search for water—a signal that they are ready to begin their northward trip to feed and lay their eggs. Although the butterflies depart as huge flocks, by the time they reach Texas they are flying solo or in small groups. Some of the original migrants survive the return trip, but it is usually the second or third generations, bred en route, that arrive at the northernmost regions in the summer.

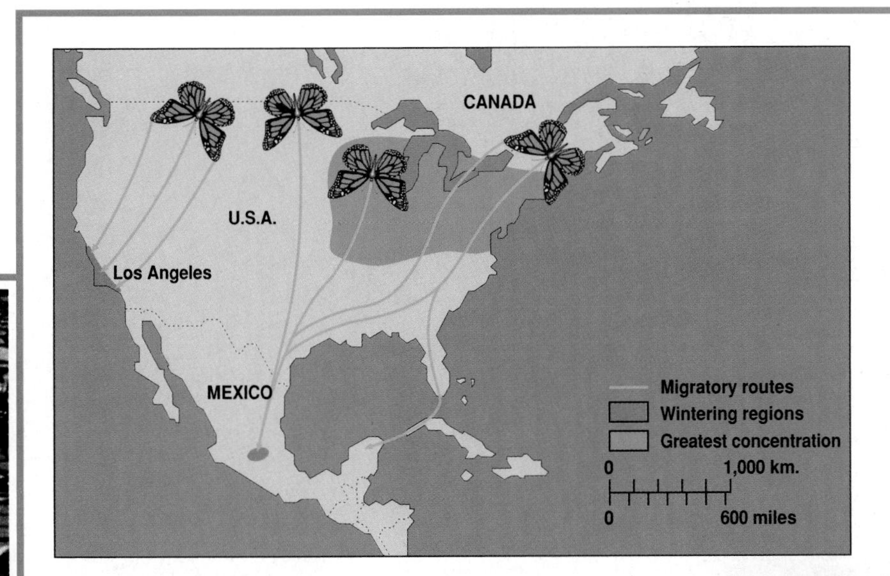

Flight plan of the monarchs
Western monarchs winter on the coast of California from San Francisco to Santa Barbara, returning in February to California's central valley and the cooler Pacific Northwest and central Rockies. Their more-numerous eastern relatives travel south in the fall from the Great Lakes region into south-central Texas and across the Gulf of Mexico to the Sierra Madre in Mexico. Some of the eastern monarchs fly down the Florida peninsula and may cross to Cuba, Mexico's Yucatan Peninsula and Central America.

After years of unsuccessful attempts at developing a way to tag monarchs, the solution came in the use of pressure adhesive labels, similar to the price tags found on glass bottles at the supermarket. At a hundredth of a gram, the weight of the tag is about one-fortieth the weight of the butterfly.

Artificial horizon
This instrument *(top)*, also called the attitude indicator, represents the airplane's attitude in relation to the Earth. An artificial horizon is kept horizontal by the gyroscope to which it is mounted. A fixed bar represents the plane's wings. As the plane climbs or dives, the bar rises above or dips below the horizon line. When the plane banks, the bar banks accordingly.

Vertical speed indicator
The vertical speed indicator *(right)*, shows the rate at which the airplane is climbing or descending (as distinct from airspeed). The instrument case contains an airtight compartment housing a diaphragm. The case and the diaphragm compartment each connect to the plane's static pressure system, and can thus register two different pressures. In level flight, the two pressures are equal and the needle points to zero. As the plane ascends, the diaphragm pressure drops faster than the case pressure and the relative difference translates into the aircraft's rate of climb.

Heading indicator
The airplane symbol on the dial *(bottom)*, reflects the actual direction of the aircraft. Should the pilot change course, the compass card underneath will show the new heading. While the compass appears to rotate, its motion is only apparent. It is the airplane that turns and the compass that remains steady in relation to the Earth.

Other natural fliers are endowed with wonderfully accurate systems for long-range navigation. A homing pigeon, set free in unfamiliar territory as far as 600 miles from its home base, returns without hesitation in a single day's journey. Other birds migrate for weeks to far-distant feeding or mating grounds, often traveling thousands of miles over featureless desert or ocean. Arctic terns are the champion long-distance fliers. The annual round-trip flight from their summertime courtship area in the far north to their winter home on the South Polar icecap and back roughly equals the Earth's circumference. To find favorable winds, they may fly at the cold, oxygen-starved altitude of 21,000 feet.

At present there is no adequate explanation for how migratory birds are able to find their way. Some scientists suggest that they atune their flight paths to the positions of the Sun and stars, calibrating daily and seasonal changes by a mysterious internal chronograph. Other researchers believe that the birds may be guided by the Earth's magnetic or gravitational fields, by means of a genetic compass passed from one generation to the next.

BASIC INFORMATION

Whether the aircraft is a single-engine Cessna or the Boeing 747 shown above, six primary flight instruments form a basic T configuration on the flight panel.

Airspeed indicator
This gauge *(left)*, is connected to a pressure-sensing Pitot-static tube and a flexible diaphragm within the instrument case. The diaphragm compares the difference between the static air pressure on the tube and the rush of oncoming air, called ram pressure. An internal lever system translates this difference in pressure into air speed, expressed in nautical miles per hour, or knots.

Turn-and-bank indicator
This instrument *(middle)*, sometimes called the needle and ball, guides a pilot through banking turns. When the needle is centered, the plane is flying straight ahead. A deflection to either side shows that the plane is turning in that direction. The ball, in a sealed glass tube filled with kerosene, works on the same principle as a carpenter's level. When the ball is centered during a turn, the plane is in balance. But if the ball rolls in the direction of the turn, it shows that the planeis side-slipping. If the ball moves away from the turn, the plane is yawing.

Altimeter
The altimeter *(right)*, is in effect a barometer that senses the decrease in air pressure that accompanies an increase in altitude. A needle linked to the mechanism indicates the height of the plane relative to the sea or the ground over which it is flying.

To match these navigational feats, the pilots of airplanes make use of complex gauges, sensors and other flight instruments that permit them to fly in virtually all weather, and at all hours of day or night. Some planes are equipped with radar, the electronic equivalent of the bats' sonic detection system. And all pilots rely on a worldwide communications network that sends signals from the ground to guide them through today's crowded skies.

THE FLIGHT DECK

Flying blind is probably the proudest skill of pilots. It means being able to maneuver in clouds or fog when nothing can be seen from the cockpit windows but grey vapor—no horizon, no ground, no sunlight, no stars. Birds, who ought to know everything there is to know about flying, cannot fly blind. Over half a century ago, an army pilot blindfolded a pigeon and threw it out of an airplane. The pigeon tried all sorts of maneuvers and then went into a spiral dive. It simply gave up. It let itself fall, tilting its wings to brake its descent. In short, it bailed out.

A pilot can fly blind, however, with the aid of artificial senses—the instruments on the panel in front of him. He must learn to put complete faith in them. Under no circumstances must he react to a bodily sensation, however strong, particularly if it disagrees with what the instruments tell him.

The human body can sense almost all the motions a plane makes. Slip or slide, climb or descent, speed-up or slow-down—all these travel through an experienced pilot's nervous system like an electric shock. What he feels corresponds exactly with what his instruments report. But there is one maneuver his body cannot detect: a banked turn. That is because the combination of bank and turn keeps the plane in perfect balance. No matter how extreme the maneuver, the bank kills the feel of the turn, and the turn erases the sensation of the bank. The turn indicator on the instrument panel shows what the plane is doing, but the pilot's instincts tell him he is flying straight and level.

This paradox explains why early aviators became utterly confused when flying through clouds or fog. Unable to see the ground or the horizon, a pilot might veer to one side without realizing it. A banking plane tends to dip its nose and pick up speed, motions the pilot would quickly sense. To correct them, he would pull back on the stick—with often fatal results. The plane's nose would be forced more tightly into the turn, the bank would steepen, speed would increase even more, and the pilot would plummet in a screaming spiral spin.

It did not take many such mishaps for aviators to realize that without special instruments to guide them, controlled flight in murky conditions was next to impossible. Today, commercial aircraft can fly at night or above the clouds, at speeds

THE WORLD AS A SPINNING TOP

First observed by Isaac Newton in the 18th Century, gyroscopic forces have amazed man for hundreds of years. The gyroscope is one of the oldest mechanical devices; in fact, the Earth itself is a gyroscope, having been thrown off by the Sun and set spinning on its axis.

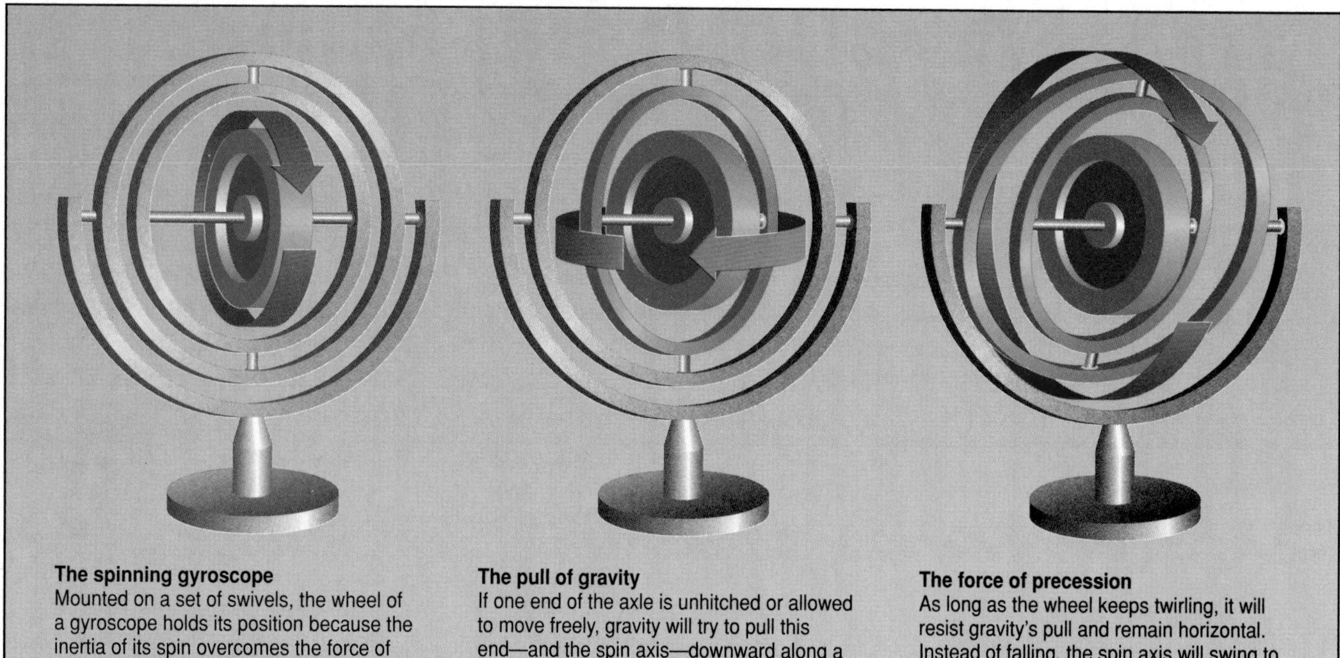

The spinning gyroscope
Mounted on a set of swivels, the wheel of a gyroscope holds its position because the inertia of its spin overcomes the force of gravity. In this example the wheel is set spinning so that its axle is horizontal; the axle corresponds to the gyroscope's spin axis. The base can be tilted in any direction, but the spin axis will remain horizontal.

The pull of gravity
If one end of the axle is unhitched or allowed to move freely, gravity will try to pull this end—and the spin axis—downward along a second axis, the so-called gravitational axis.

The force of precession
As long as the wheel keeps twirling, it will resist gravity's pull and remain horizontal. Instead of falling, the spin axis will swing to the side—a phenomenon called precession. When an airplane maneuvers, the force of precession stabilizes such gyroscopic instruments as the artificial horizon, the heading indicator and the turn-and-bank indicator.

PRESSURE, SPEED AND ALTITUDE

At the heart of flight instruments that measure altitude, speed, and rate of climb is a Pitot-static tube. A small, open-ended pipe installed on the nose or wing, the tube measures the difference between two pressure values. The airspeed indicator at right compares the static pressure of the atmosphere against the impact, or ram pressure, of the oncoming airstream.

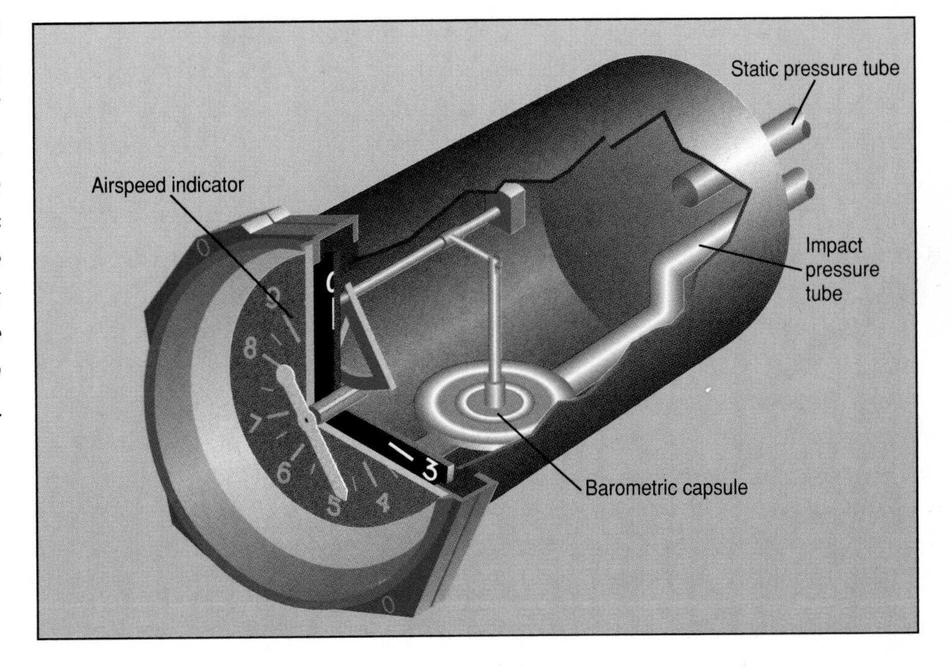

exceeding 600 miles an hour, often never within sight of the ground from takeoff to landing. The pilot flies entirely by his artificial senses, relying on the battery of alert and precise instruments in front of him. In the sweep of aviation's progress, this is perhaps the biggest accomplishment since the Wright Brothers.

The modern flight deck is a bewildering array of dials, gauges, switches, levers and buttons. The instruments fall into three broad categories. Systems instruments provide detailed information about the aircraft itself—its engines, fuel system, landing gear, flaps and control surfaces, and much else besides. Navigation instruments and equipment allow the pilot to establish his course and monitor his position. The third group—and the most important one for flying blind—depicts the moment-to-moment relationship between the aircraft and the ground below.

In virtually all planes, the "blind flying panel" contains six primary flight instruments: an altimeter, an airspeed indicator, a vertical airspeed indicator that shows the rapidity of ascent or descent, a heading indicator that tells where the nose of the plane is pointed, a turn-and-bank indicator, and an attitude indicator that serves in effect as an artificial horizon. The instruments can be displayed as cockpit dials or digital displays on a composite instrument screen.

Mounted either on the airplane's nose or near the leading edge of a wing, a so-called Pitot-static tube measures two kinds of air pressure and compares the difference. As the plane moves forward, the tube's front end catches the oncoming airflow and registers its force; this information, relayed to the instrument panel, will give the pilot his airspeed. At the same time, holes in the side of the tube measure the local atmospheric pressure—or, more precisely, the pressure the air exerts on the tube if both were motionless, or static. The static pressure serves as a benchmark for calculating the oncoming air pressure. It also provides the data for two other pilot instruments. Since air pressure naturally diminishes with height above sea level, the Pitot-static tube can determine the plane's altitude. And, when the altitude changes during a climb or a dive, the rate of change appears on the dial of the vertical airspeed indicator.

A few pre-flight adjustments are needed to ensure the instruments' accuracy. Since the prevailing atmospheric pressure varies with local weather conditions, this figure must be noted and the gauges adjusted. A correction must also be made for the air temperature. Cold air exerts more pressure than hot, and differences in local thermometer levels can skew altitude readings by hundreds of feet.

The other primary flight instruments—the heading indicator, the artificial horizon and the turn-and-bank indicator—rely on the marvelous ability of gyroscopes to hold their positions no matter what. At one time, gyroscopes belonged to a branch of mechanics of interest mostly to children: A spinning top does not topple over and a coin rolling across the floor stays upright on its edge. The Earth itself is a gyroscope, steadily spinning on its axis in space. Thanks to the inertia of its rotation, a gyroscope resists any effort to change its bearings. Mounted on swivels behind a plane's control panel, it is a center of stability, maintaining its level-headed attitude through the sharpest turns, rolls, dives and climbs. It becomes in effect a miniature, on-board representative of the stable Earth below.

The aviation genius who first put gyroscopes into the air was Lawrence Sperry, son of the founder of the Sperry Gyroscope Company and the American pioneer of blind flying. He designed a turn indicator that allowed pilots to navigate through heavy, low-lying clouds, make banked turns, and come out flying straight and level. His Sperry Artificial Horizon, developed in 1929, made visual reference to the

MULTI-TIERED SKYWAYS

Aerial superhighways crisscross the sky in regions of controlled airspace, where all planes are subject to compulsory air traffic control. Low Level Airspace extends to about 18,000 feet above sea level, where one set of traffic rules prevails. Another set governs High Level Airspace, which starts at 18,000 feet and is the domain of jets.

real horizon unnecessary. Although refined, it is still in use. The instrument's dial, connected to a gyroscope spinning at several thousand revolutions per minute, shows a line representing the horizon and a bar that mimics the position of the airplane's wings. When the plane banks, so does the wing bar. When it climbs or dives, the wing bar moves up or down in relation to the horizon line.

Another Sperry innovation was the Directional Gyro, a kind of non-magnetic compass that shows the plane's heading. On the dial's face is the image of a miniature airplane, superimposed above a freely-moving compass card. While on the ground, the pilot adjusts the card to agree with his conventional on-board magnetic compass; a spinning gyroscope locks in this information. In flight, the gyroscope holds its course, even when the pilot veers left or right. The card appears to swing, and the nose of the miniature plane points to the new heading.

Gyroscopes also play a critical part in two sophisticated computer systems. One is the autopilot, which takes over the routine control corrections needed for stable flight. A pair of gyroscopes, one spinning horizontally and the other vertically, detect any deviation from a set heading. A change in the plane's attitude or course activates small motors that make adjustments to the various control surfaces.

The other device is a navigational tool called the internal guidance system, or IGS. At its heart is a set of motion detectors, or accelerometers, that track the slightest changes in momentum. Should the plane accelerate in any direction—up or down, fore or aft, left or right—the accelerometers will respond. The role of the gyroscopes is to steady these devices while they take their readings. A continuous stream of data feeds from the accelerometers into the IGS computer, which uses it to calculate the plane's position. Even in the densest fogs and blackest nights, a pilot guided by an IGS always knows exactly where he is.

A PILOT'S SKY MAP

To an experienced pilot, this jumble of lines, numbers and codings indicates the normal traffic patterns between airports. Ground stations are designated as circles; they provide the pilot with vital navigation information en route.

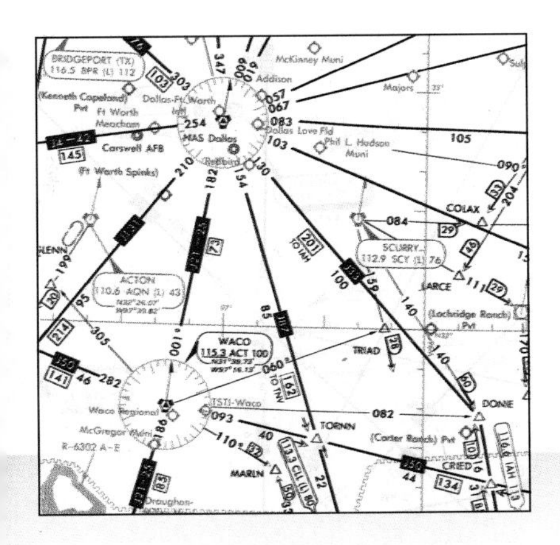

HIGHWAYS OF THE SKY

With so many airplanes in the sky today, a problem pilots face is keeping out of each other's way. Thus, a widespread system of invisible aerial highways and superhighways runs through congested areas to make this easier. Pilots travel along numbered air routes, or corridors, in a three-dimensional traffic network. A corridor is about nine miles wide and, at lower altitudes, 1,000 feet above or below any other; at levels beyond 29,000 feet, the separation expands to 2,000 feet. As a pilot flies along his designated corridor, he follows well-established rules of the air, just as an automobile driver respects local traffic codes. Interlocking radio signals, beamed from ground stations, direct the traffic and mark the intersections: his signposts. The transmissions are marked on charts that the pilot uses to track his position—the way a motorist takes his cue from road maps and road signs.

The system is a far cry from the techniques of aviation's early decades. Back in the 1920s, even airmail pilots tackled navigation the hard way: eye contact with the ground at any price. With little more than a clock, a compass and a map, they flew through good weather and bad, keeping a close lookout for landmarks below. They would try to slip under dense clouds and fog, often hedgehopping just above the treetops and using railroad tracks as guides. In those heroic days, bonfires marked the hills along the mail routes and helped serve as checkpoints. On clear nights, the pilots would fly by the stars, using an aeronautical version of the mariner's sextant and a chronometer to determine their positions.

Small private planes still fly at low altitudes and only under conditions of acceptable visibility, when the pilots can eyeball the ground and any other traffic in the vicinity. They operate under a convention of basic traffic ordinances known as visual flight rules (VFR). Most larger aircraft, and commercial jetliners in particular, now travel "on instruments" at all times, their pilots visually oblivious to the ground below. They observe a set of internationally-enforced instrument flight rules (IFR) and follow radio instructions from air traffic controllers along their route.

Air traffic controllers are the ground-based traffic cops of the airways. From the moment a jetliner pulls away from the terminal at one airport until it taxis to a stop at another, it is kept a safe distance from other planes by men and women on the ground. Every airplane, wherever in the world it may be, is in radio hailing distance of one or more controllers, who monitor the skies in their vicinity by means of radar. A microwave transmitter at the control center sends out radio signals on frequencies above those used in broadcasting, sweeping them across the sky in a 360-degree arc. The signals bounce off any airplanes they hit and return to the center; the radar echo from an airplane 200 miles away speeds back in only one-five-hundredth of a second. Large commercial and military aircraft carry a transponder in their bellies; triggered by the ground signal, it transmits a characteristic return signal that identifies the particular aircraft. The reflected signals, picked up by the center's antennae, show up as a blip on the traffic controller's computerized radar screen. The controller uses the tagged information to identify the plane, along with its precise location, altitude and ground speed.

A bird's-eye view of New York's John F. Kennedy Airport belies the frenzied activity behind the scenes—cleaning, inspecting, servicing, refueling, loading and unloading an average 400 airplanes per day; and boarding and disembarking some 80,000 passengers. It is the responsibility of ground controllers to make sure that the planes taxiing to and from the gates do not collide with each other, or with any of the dozens of service vehicles buzzing about.

Like a line of cars waiting for the light to change, commercial jets queue up to await final takeoff clearance from the airport control tower. Once each plane is airborne and passes beyond local airspace, the controller will hand it off to the next control center along its route.

At many airports, trained falcons such as this one keep runways clear of pigeons and gulls. Otherwise, the birds might be sucked into aircraft turbines or propellers, causing expensive damage and possibly engine failure. The falcons make few kills; their job is to scare the intruders away.

TWA
TRANS WORLD
IBERIA
KLM

In heavily trafficked sectors, a controller must use all his skills and experience to convert the multitude of two-dimensional images on his screen into a three-dimensional picture of aircraft activity. And he must constantly think in a vital fourth dimension: that of time. For all of this air traffic is moving at different speeds, and he must respond quickly to divert any planes that are on collision courses. He alerts the pilots by radio and orders the necessary changes in their flying speed, altitude and direction.

Virtually every move that a large commercial jet makes, from the time it takes off until it lands at its destination, is dictated by air traffic controllers. The controllers are "pilots on the ground," and each segment of the flight comes under the supervision of one of four groups of them. The first group, the surface or ground controllers, directs the taxiing of planes to and from their proper terminals and gates. Local airport controllers govern takeoffs and landings; the bubble of space under their jurisdiction may reach upward as high as 8,000 feet and cover a diameter of more than 20 miles. Farther out, approach and departure controllers take responsibility for aircraft flying within 25 to 60 miles of the airport. As the plane speeds on its journey, its supervision is handed over to a series of en route controllers along its flight corridor.

In the United States, this system comprises 20 en route control centers dotted about the countryside, more than 400 airport towers and some 15,000 controllers. When air traffic is light, each controller may be responsible for only one or two aircraft. But during peak hours at busy hubs, like Chicago's O'Hare or Los Angeles International Airport, the number can increase to 15. Keeping so many planes on smooth courses requires cool nerves and split-second timing. One controller, asked what kind of people are good at his work, replied: "Not the ones who have trouble making up their mind."

FROM DEPARTURE GATE TO TAKEOFF

For a typical commercial flight, traffic control begins at least half an hour before takeoff, when the pilot files his flight plan with the departure flight planning office. The plan includes the aircraft's call sign, the type of aircraft, the requested route and altitude, the proposed departure time, the estimated flying time from takeoff to arrival, the amount of fuel on board and the type of navigation equipment being used. The flight planning office then alerts the airport's traffic controller, the depar-

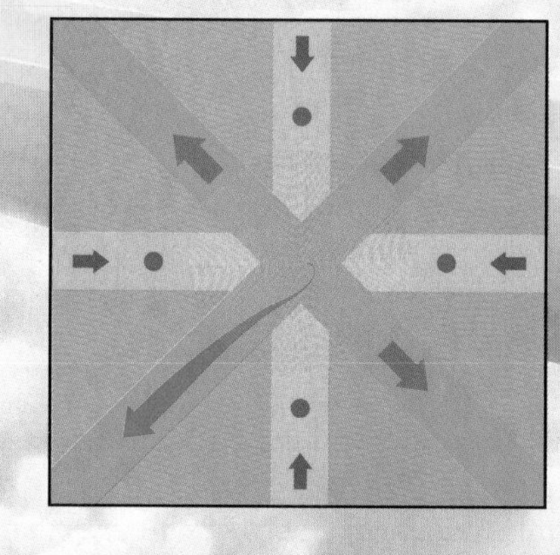

BEACONS AND BEDPOSTS

After takeoff, a commercial jetliner climbs along a pre-assigned access route to its cruising altitude, where it will level off. The pilot is guided by a set of four radio beacons, each situated about 40 to 45 miles from the control tower. Viewed from above, the beacons surrounding an airport resemble a four-poster bed (inset). *Incoming planes fly directly over a bedpost as they descend to the runway; departing planes generally fly a course between two beacons as they ascend to their respective flight levels. Departure controllers track the plane until it attains its proper level, then hand off radar surveillance to the next controller en route.*

TRACKING THE PLANE FROM THE GROUND

A transponder—standard equipment on all commercial airlines—transmits the plane's identification code, altitude, course and speed. This information appears as a data block of numbers and letters on the controller's radar screen.

ture controller and the en route controllers in the sectors that the plane will be flying through.

After the flight plan is approved, the pilot radios the airport control tower and receives his first instructions from the clearance-delivery controller, part of ground control. These instructions include the plane's takeoff position, the radio frequency it will use on departure, its transponder code, and any changes in the flight plan, such as delays caused by congestion. The pilot then pulls away from the gate and taxis to his designated runway. A runway number painted on the tarmac indicates the direction the plane will head as it takes off. For example, runway number nine means that the plane will point in a compass heading of 90 degrees, or due east. If two runways are laid out alongside each other, one is designated "right" and the other "left."

When the plane reaches the end of its runway and is number one for takeoff, a local airport controller takes over from ground control. From his post in the tower, the controller has a 360-degree, bird's-eye view of the scene below and the sky above; only when inclement weather obscures the runway does he resort to radar. But whatever the conditions, he knows where the plane is at all times. With a final check of the runway, he clears the flight for takeoff. The engines roar to full power and the jet accelerates down the runway.

As the plane reaches takeoff speed, the pilot pulls back on the control column and points the nose skyward. Once airborne, responsibility for the aircraft passes from the airport tower to a departure controller, who should now be able to spot the plane as a coded blip on his radar screen. Receiving radio confirmation of this handoff, the pilot heads out on course. Within five to ten minutes, responsibility for the flight passes from the departure controller to the first in a succession of en route controllers.

TRANSATLANTIC RELAY

A plane crossing the Atlantic moves along a variable succession of corridors that are selected every 12 hours from an internationally published track list. The choice depends primarily on weather and wind conditions en route. One critical factor is the position of the jet stream. A migrating current of fast-moving westerly winds at great altitude, the jet stream can add or subtract as much as one hundred miles per hour from an aircraft's overall speed, depending on whether the plane is riding with it or bucking it. The flyways are identified alphabetically: Track Alpha, Track Bravo, Track Charlie, and so on. Over the Atlantic, they are 60 nautical miles wide and the same distance apart; the aircraft traversing them must be separated by a minimum of ten minutes in time and 2,000 feet in altitude.

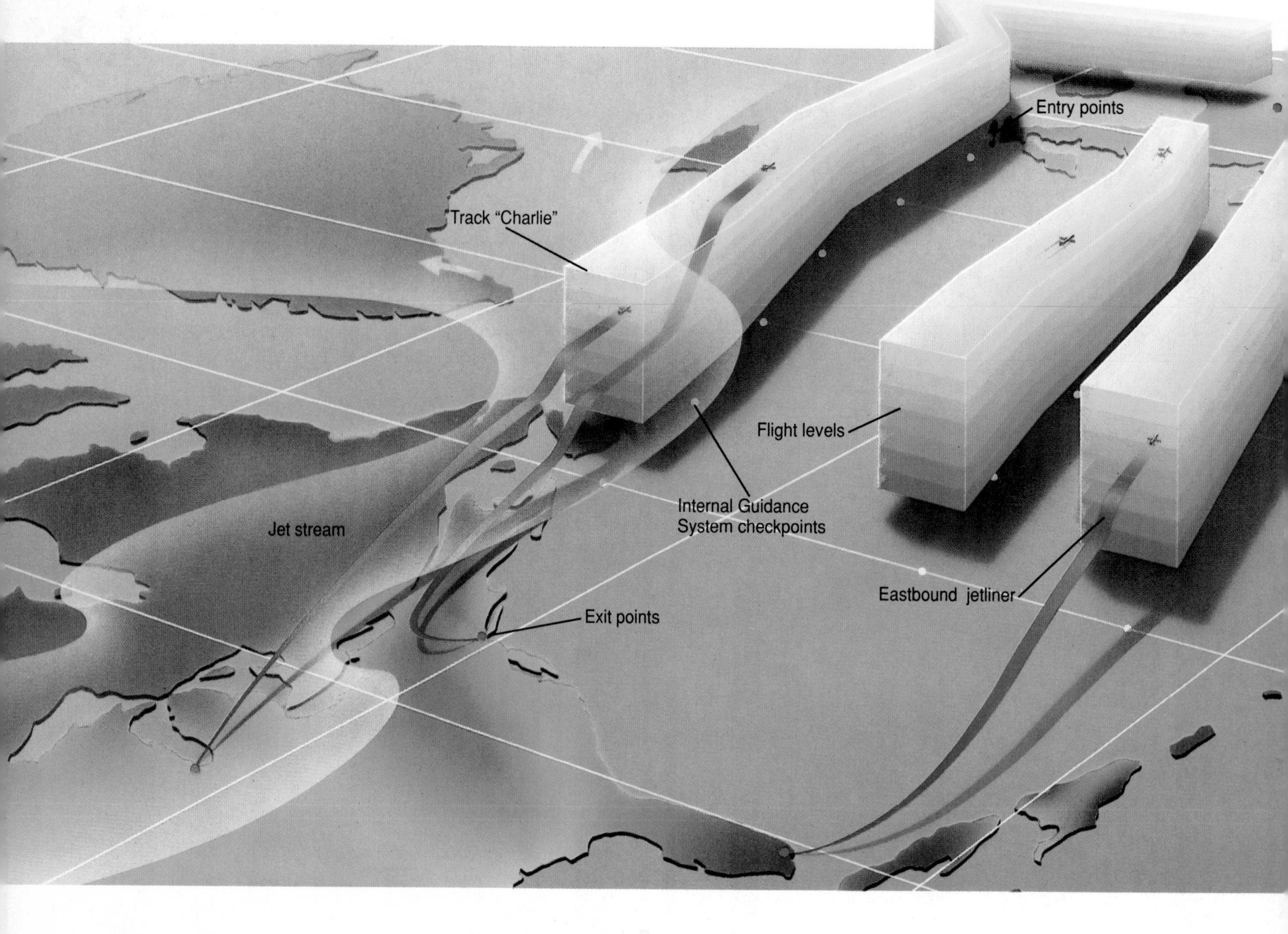

This particular flight from New York's John F. Kennedy Airport is destined for Rhein-Main Airport in Frankfurt, West Germany. When filing the flight plan before takeoff, the pilot requested a preferred corridor—Track Charlie, as it happens—along with a desired airspeed of Mach .85 and a projected altitude of 35,000 feet. The plan approved, he notes down the latitude and longitude coordinates of the geographic checkpoints he will be passing. As he reaches each one, he will report his altitude and position to the nearest traffic control center—which will also be monitoring the plane's progress on its radar screens.

The initial portion of the journey is over land. As the plane flies from checkpoint to checkpoint, it is guided by a series of VOR radio beacons, or Very High Frequency Omnidirectional Range beacons, that broadcast at static-free wavelengths just above normal FM household radio signals. As the plane travels, it is methodically handed off from one controller to the next. In North American air space, the control centers are typically 300 to 400 miles apart; a flight from New York to Frankfurt passes through four en route centers between the two airports. Each radar handoff represents a critical step along the way. Until a control center establishes contact with the next center on the plane's route and the handoff is accomplished, the plane is not allowed to proceed. If necessary, the pilot adjusts his flight plan on instructions from the ground.

The relay continues until the flight nears Gander, Newfoundland. By this time it has reached its designated cruising altitude of 35,000 feet. Gander Oceanic Control reconfirms the flight plan, or calls for an adjustment based on weather or traffic. The plane moves out over the North Atlantic. The pilot now activates the on-board Internal Guidance System (IGS). He has already punched in the appropriate transatlantic coordinates back at the departure gate, in anticipation of the flight's transoceanic leg. The IGS computer takes charge, noting each motion the plane makes and comparing it with earlier data to calculate the ongoing position. Should wind or weather jog the plane off course, the IGS, together with the autopilot, will make instantaneous corrections.

The IGS checkpoints are located at intervals of 10 degrees longitude; the plane flies unerringly to each one. At each point, the pilot radios his position to the appropriate control center. For the first third of the ocean passage he reports to Gander. His next reports go to Reykjavik, Iceland. Finally, as he nears Europe, he communicates with Shanwick Ocean Control in Ireland. Shanwick governs his movements until the flight makes its land entry over Cork, where it comes under the control of London Center. From this position on, the flight follows standard traffic procedures for Europe. Fortunately for American pilots, English is the international language of the air, and there is no difficulty in communicating with controllers in Frankfurt for the final approach and landing.

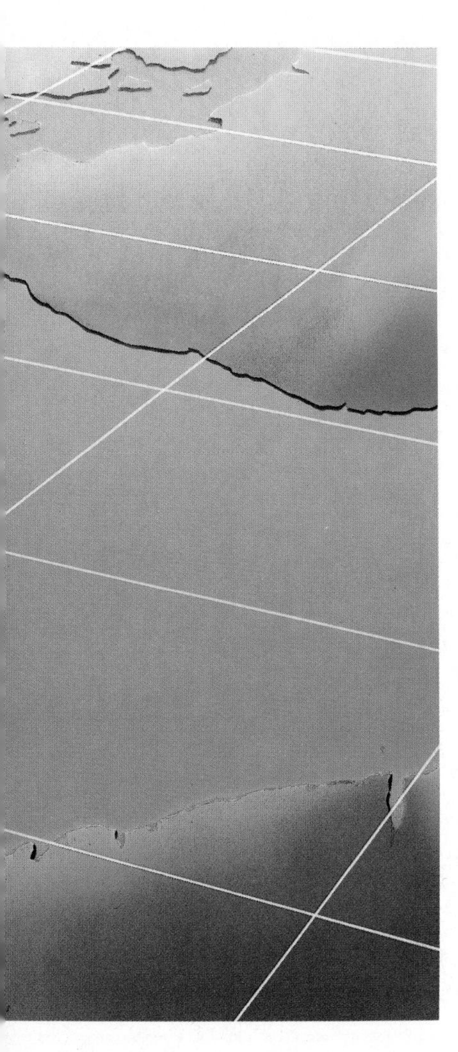

RIDING THE JET STREAM

Flying on Track Charlie, one of many east-west transatlantic air corridors, a commercial jetliner enters the jet stream, about 100 miles wide and some two miles from top to bottom. With the westerly stream at his back, a pilot can cut his travel time from North America to Europe by as much as one hour.

AVOIDING MIDAIR COLLISION

With the dramatic growth in commercial aviation, midair collisions present a very real peril; the definition of a near miss, meanwhile, varies with altitude. And they are not entirely modern problems. The first documented midair collision occurred over Milan, Italy, in 1910, during an air show. A Frenchman, piloting a monoplane, plowed into a biplane flown by an Englishman at a then breakneck speed of 50 miles per hour. Both survived the crash.

Since then, given the rapidity of growth in the number of air miles flown each year, it is remarkable that such disasters do not occur more frequently. It is statistically safer to fly in an airplane than to travel by automobile. But near misses have more than doubled in the past decade, and when collisions do happen, the heavy toll in human life can be devastating.

On the bright, sunny morning of September 25, 1978, Pacific Southwest Airlines Flight 182 was nearing Lindbergh Field at San Diego, California, at the end of its regular commuter run from Sacramento. The 66-ton Boeing 727 carried 128 passengers and seven crew, including flight attendants. The pilot, heading due east, was awaiting instructions from approach control to make the 180-degree turn that would put him on a direct path to his designated runway, number 27. At the same time, a two-seater Cessna 172 on a pilot-training exercise had just skimmed the same runway in a simulated landing maneuver. It too was flying east, and rapidly gaining altitude. Despite the day's brightness, visibility for both pilots was partly obscured. The Boeing pilot, flying level, could not see beneath him. The pilot of

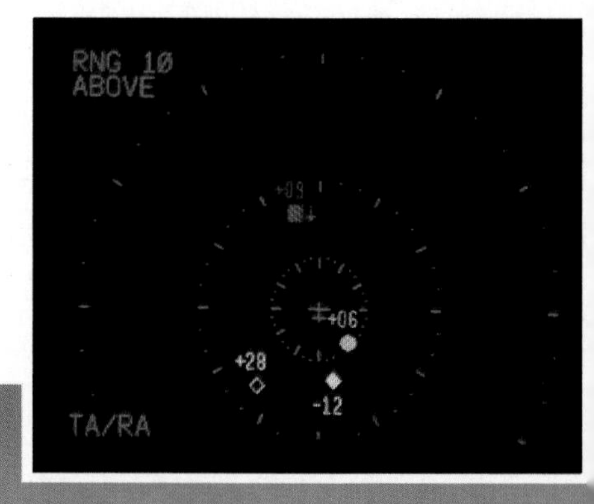

the Cessna had a blind spot overhead, where the plane's high-placed wing cut off his line of sight. Both had the sun in their eyes. And so they collided.

The Boeing hooked the Cessna with its nose wheel, and the two planes plunged to Earth. The horrified traffic controller, unable to divert either plane in time, watched the falling wreckage on his radar sceen. "An aluminum shower," is how he described it. The chunks of flaming metal fell into a residential area, setting fire to ten houses. The death toll: all 137 occupants of both planes and 13 San Diego suburbanites. It was the worst air tragedy in North American aviation history.

The air fatality count over Europe has been as equally heartrending. For example, in September 1976, a British Airways Trident collided with a Yugoslav Air DC-9 in the skies near Zagreb; all 176 people aboard the planes perished. To head off such air disasters, traffic controllers direct every effort to keeping planes a safe distance apart.

Flight rules internationally have been tightened, and many commercial jets are equipped with a sophisticated, computerized radar that serves as a backup to the traffic controllers on the ground in their efforts at separation. Called the Traffic Alert and Collision Avoidance System (TCAS), it reinforces the "see and avoid" capability of the crew on the aircraft. A TCAS, which includes an advanced type of transponder, can track as many as 24 other transponder-equipped aircraft within a five-mile radius, evaluate collision potential, and recommend the proper evasive action for almost any flight circumstance. If a convergence seems imminent, it warns the pilot at least 40 to 45 seconds beforehand.

A TCAS-equipped aircraft is in effect surrounded by three concentric envelopes of radar-monitored airspace. The envelopes are defined by flight time, so that the extent of each one depends on the combined speeds of the aircraft involved. The space immediately around the aircraft is the collision area. Next, there is a warning area that represents 20 to 25 seconds distance in closing time between the collision area and an intruding plane. The largest envelope, the caution area, expands outward to mark the critical 45 seconds of response time.

These seconds are enough time to avoid aerial catastrophe. When an intruder enters a pilot's caution area, his radar immediately picks it up and displays its relative bearing and altitude on the screen. Should the intruder move into the warning area, the TCAS issues a series of flashes and beeps, along with specific instructions for averting a collision. Traffic controllers, monitoring the two planes from their posts on the ground, make the final decision as to which plane should change its flying altitude or course.

INTRUDER ALERT

The two airplanes in the diagram have diverged from their assigned air corridors and are flying on a collision course. Warned by his TCAS, the pilot picks up the intruder on his radar screen (inset). *The screen shows the pilot's position as a jet symbol surrounded by a ring of dots. The intruder can be seen near the top of the screen. A blue diamond would indicate no threat; an amber circle represents a moderate threat; and a red square would mean an immediate threat that calls for quick evasive action. In this case, a synthesized voice alerts the crew to ascend to a higher altitude, while the other craft maintains its course.*

THE PROBLEM OF WEATHER

Fog, snow and rain are unavoidable conditions that pilots are trained to fly through. All restrict visibility and, at the very least, they add to the difficulties of traffic control and cause expensive delays in flight schedules. One of the worst weather problems is freezing rain, which may fall even in summertime at high altitudes. Consequently, every jetliner is equipped with deicers, electric heating coils embedded in the windscreen, and deicing fluid systems that protect the leading edges of the wings. An ice deposit of just half an inch in thickness on the leading edge of a wing can reduce its lifting power by as much as 50 percent and increase drag by an equal amount.

One weather phenomenon that pilots attempt to avoid at all costs is a thunderstorm. The danger of a thunderstorm to them comes from three potential killers—high turbulence, hail and lightning.

All thunderstorms begin as unthreatening cumulus clouds. But an atmospheric imbalance can quickly transform these harmless-looking puffs into gigantic storm factories capable of producing every type of hazardous weather condition known to aviation. Their turbulence can bounce a plane like an ice cube in a cocktail shaker, severely stressing wings and fuselage. The hail often associated with them has been known to pummel down in chunks the size of golf balls. Even smaller particles can shatter a plane's windows, dent its wings, and dislodge the cover that protects its sensitive radar antenna.

A lightning strike can wreak absolute havoc. Packing enough electricity to light up ten city blocks, the average bolt can generate temperatures of 18,000° F., or twice the surface temperature of the Sun. It can punch holes in a plane's aluminum skin, damage engines, and destroy communication and navigation equipment. Lightning has been known to weld the hinges of ailerons, and momentarily blind pilots with its intense flash. And while it seldom harms anyone in the cabin of an aircraft, on occasion it has sheared off sections of the wings and tails from both light planes and heavy transports.

A pilot's first line of protection is the weather report, and whenever possible he plots his course to avoid likely storm areas. In flight, he constantly monitors

A SLICE OF THE SKY

A massive and menacing storm looms ahead of a commercial airliner. Using on-board weather radar, the pilot compares the instrument's cross-sectional profile of the storm to the view from his windscreen. Following the directions of the weather radar, he turns left to a compass heading of 100 degrees, while the storm moves slowly past him from left to right.

the sky for the anvil-shaped cumulus clouds that indicate thunderheads and he gives them a wide berth. The storms build rapidly and they typically cover many miles of airspace. Lightning can occur anywhere in their vicinity, striking from cloud to ground, cloud to cloud, or at any large passing object. Some discharges have streaked as far as 90 miles from their source. And an airplane, with its sharp, flat metallic surfaces, makes an ideal lightning rod.

All commercial airliners carry on-board weather radar to help them stay out of harm's way. Heavy precipitation sends back a strong radar echo, which shows up on a computerized, color-coded radar screen. Rapidly shifting colors indicate fluctuations in rainfall and possible turbulence or hail. The more advanced radars can pinpoint areas where lightning strokes are concentrated.

Radar is not infallible, however. Since its beams reflect only off moisture, it cannot detect fair-weather turbulence, or even hail, unless the ice crystals are coated with rainwater (nor can it distinguish between rain and wet hail). Furthermore, an airplane's most hazardous moments come not in the center of a storm nor at its height. Studies of lightning strikes show that a plane is more likely to be hit when a storm is dissipating, and when turbulence is low and rainfall light. Under these conditions, the air may still carry a potent charge of static electricity, which is triggered only when the plane flies through it. All things considered, a pilot's best strategy is to skirt the storm area entirely and radio the nearest ground controller for permission to change course.

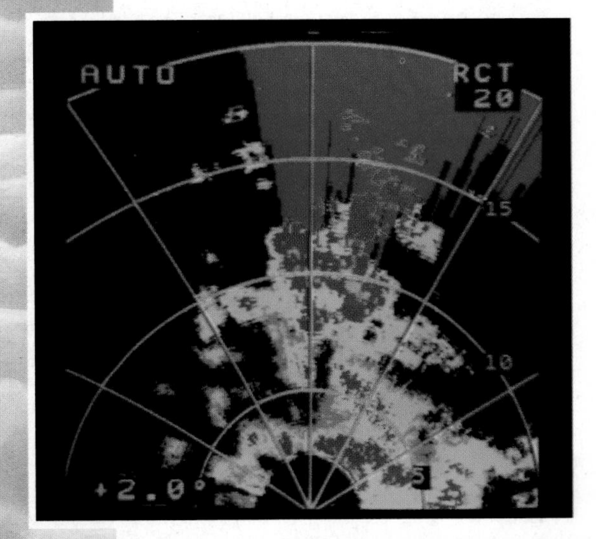

The wide-angle view of a weather radar screen (below) *shows two storm cells 10 to 12 miles ahead. The radar beam has electronically sliced a cross-section of each storm and the computer presents each slice as an overhead view. To target different layers, the radar operator would tilt the beam up or down. Variations in color indicate the severity of the rainfall.*

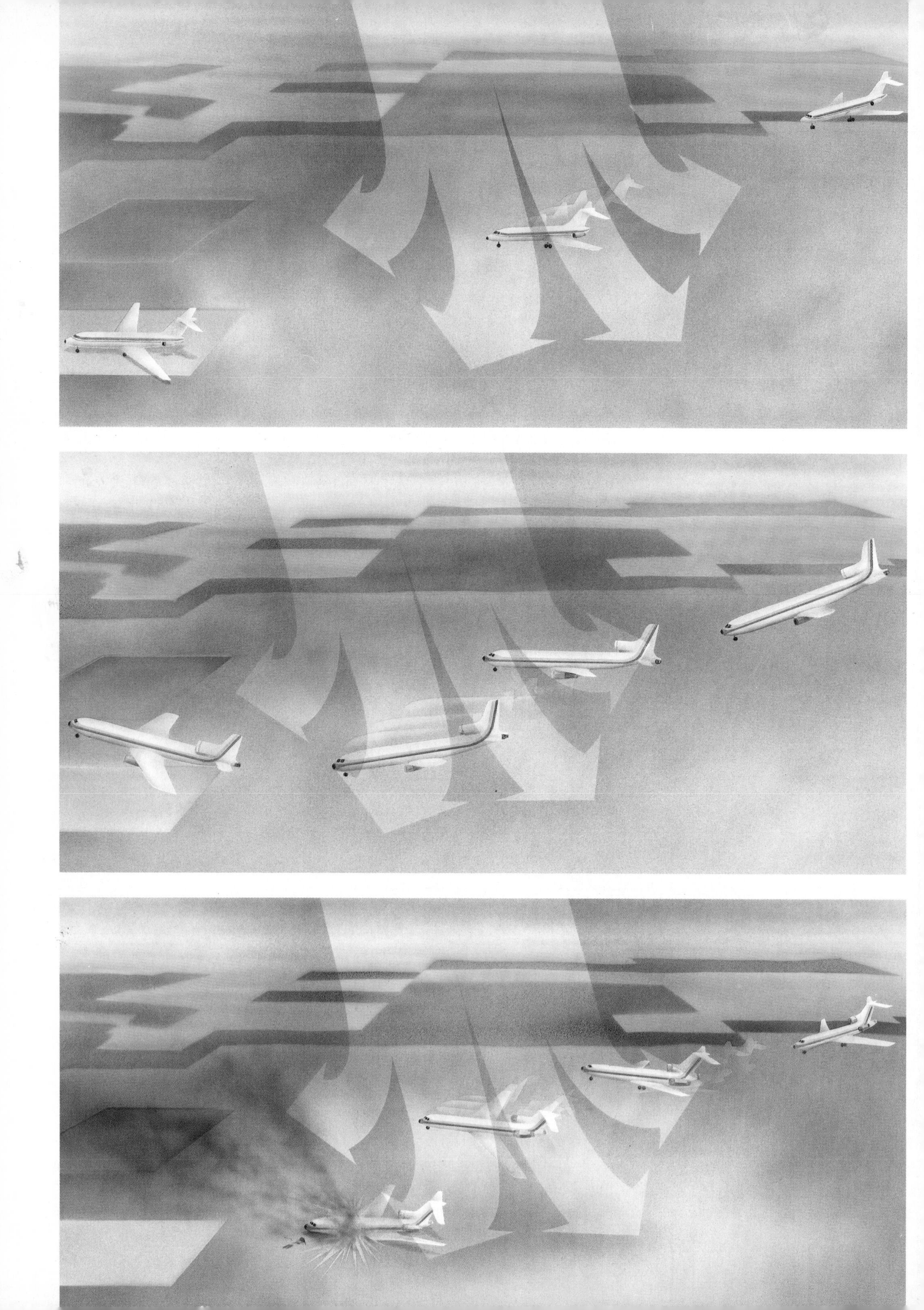

DIARY OF A DISASTER: JUNE 24, 1975

3:48 P.M.
Air traffic controllers at New York City's John F. Kennedy Airport note showers and lightning over the approach course of Runway 22 Left. Unseen are powerful downdrafts, called microbursts, that are the breeders of wind shear. Allegheny Flight 858 enters the first microburst, which apparently is weaker than those to come. Headwinds give the plane extra lift and the pilot reduces power. After passing through the microburst's central downdraft, the plane encounters lift-robbing tailwinds. To compensate, the pilot opens the throttle for maximum thrust and lands safely.

3:58 P.M.
Eastern Flight 902 suffers a drastic air-speed loss as it enters the backside of a second microburst. Reacting quickly, the pilot applies maximum power in order to abandon his approach. Even at full throttle and with the nose tipped up, the jet continues to sink—from 400 feet to 100 feet. Finally, a mere 60 feet above the ground, the aircraft levels off. The jetliner banks left and flies its shaken passengers to Newark Airport.

4:05 P.M.
At 500 feet, Eastern Flight 66 enters a rain shower and an extraordinarily strong third microburst. The captain calls for "takeoff thrust"—maximum power. But little more than a second later the jet crashes short of the runway, killing 113 people in the worst single-plane disaster in U.S. commercial aviation history to that date.

THE UPSIDE-DOWN TORNADO OF WIND SHEAR

"There are two critical points in every aerial flight—its beginning and its end." So wrote Alexander Graham Bell, inventor of the telephone, in 1906. Bell knew what he was talking about. Having turned his attention to aviation, he founded the Aerial Experiment Association, which was largely responsible for inventing the aileron. And what was true then is still true today.

As the jetliner begins its descent from 30,000 feet, the captain announces that things might get rough. The sky darkens suddenly as the plane enters the clouds and rain pelts across the window. Strong winds buffet the plane. Moments from touchdown, the engines roar to full power; a wing drops sharply, then rises. The plane breaks out of the rain, the ground rushes past and the plane is on the runway.

Many air travelers have white-knuckled their way through tense moments like these, relying on the skill of the professionals in the cockpit to deal with them. Maneuvering on or close to the ground—during takeoff or landing—is the most challenging aspect of flying. It accounts for nearly two-thirds of all fatal accidents since 1959. Fog, snow, haze and rain have contributed to more than half of these crashes. Recently another enemy has been pinpointed: low-level wind shear, or any sudden change in wind speed or direction.

Wind shear is produced by microbursts of violently moving air, sometimes referred to as upside-down tornadoes. Microbursts form in rain clouds when evaporating raindrops create a pocket of cool, heavy air that falls in a powerful downdraft. When the air hits the ground, it fans out like the contents of a bucket of water spreading across the pavement. Microbursts occur at low altitudes and are usually only one-half to two miles in diameter. They can build to maximum wind speeds in as little as two minutes, and sometimes last only five minutes. But they pack a mean wallop. Fifty were once detected near Chicago's O'Hare Airport in a span of forty-two days.

When an airplane enters a microburst, headwinds increase the speed of air flowing over the wings, giving the plane additional lift. Then, in rapid succession, the plane encounters a downdraft in the center of the microburst, followed by tailwinds that drive it toward the ground. A pilot's first reaction is to reduce power, as a motorist going downhill eases pressure on the accelerator. But as the headwind decreases, additional power is necessary. Apply full thrust? Yes, immediately, but the situation is complicated by the characteristics of a jet airplane. For all its virtues, the turbine engine does not react as fast as the old-fashioned piston engine. The plane, usually at 70 percent of full power during this stage of the approach, takes several seconds to generate maximum thrust. In that brief interval, it might lose too much altitude to pull out of its fall.

Wind shear's most severe tantrums are rare, short-lived and difficult to predict with accuracy. Following the dramatic crash in 1975 of Eastern Flight 66 at John F. Kennedy Airport in New York City *(left)*, the Federal Aviation Agency declared war on wind shear. In cooperation with the major carriers, the agency embarked on a four-year program that included the development of ground-based devices to detect wind shear and airborne equipment to help pilots combat it.

To deal with these infrequent but significant threats to air safety, commercial pilots are trained to maneuver out of microbursts, assisted by on-board detection systems that compare the plane's speed with that of the wind. Large discrepancies signal potential wind shear conditions. On the ground, airports maintain Low-Level Wind Shear Alert Systems, which have recently been upgraded by doubling the number of detectors.

CLEARED TO LAND

As an airliner nears its final destination at a typically congested big-city airport, the captain receives instructions from the local tower to enter a holding pattern—a tiered sector of the sky clearly defined by a radio signal beamed straight up from the ground. This aerial ladder is a traffic management system that keeps airplanes circling safely until a runway is free for landing. Each airplane enters the stack at the top—perhaps as high as 23,000 feet—and begins flying around an oval race-track nine miles long and four miles wide. Below it, at thousand-foot intervals, other aircraft circle the signal beacon, awaiting their turn to land. When the airplane at the bottom of the ladder is cleared to make its final approach, the plane above it takes its place. Each aircraft moves down one rung, with a new arrival entering the top position in the holding pattern.

When approach control clears a plane for landing, the pilot begins his intermediate approach, guided by a surveillance radar system. At about 3,000 feet, he then begins his final approach to the runway. Even if the runway is obscured by clouds or fog, the pilot can land the plane as long as he makes visual contact with the ground at 200 feet, and visibility on the ground is no less than a half mile. If not, the pilot must abort the landing unless he is specially certified to make a fully automatic approach.

Another set of radio beams—called the Instrument Landing System (ILS)—guides the pilot on his final approach. One is the glide-slope beam that indicates the plane's angle of descent. The second, intersecting the glide-slope beam vertically, is the localizer beam, which projects a pair of signals to the left and right of the runway centerline. Flying between them, the plane heads straight for the runway.

Inside the cockpit, the captain lowers the landing gear at about 2,000 feet and keeps his eyes on an ILS instrument with intersecting crossbars. These vertical and horizontal lines indicate the plane's relative position to the runway: left or right of the centerline and high or low on the glide path. When the lines are centered, the pilot knows his plane is descending toward the center of the runway at an angle of 2.5 degrees. Two vertical radio marker beams tell the pilot his distance from the start of the runway. The first marker is about five miles out, the second half a mile from the runway. At the second marker, the plane is about 200 feet up. Twenty seconds later it touches down, and the pilot reverses the engine's thrust and applies the brakes. Following the ground controller's directions, he taxies to the terminal building and gate.

The Instrument Landing System may ultimately be superseded by newer devices such as the Microwave Landing System (MLS). Unlike the single, fan-shaped glide path of the ILS, the MLS provides multiple entry points at varying distances from the runway to accommodate aircraft with different approach speeds. In this way, light aircraft can keep clear of dangerous vortices generated by larger, heavier planes. Given the continuing boom in air travel, increased capacity will be an important feature of tomorrow's airports; MLS may make it easier to schedule more takeoffs and landings in safety.

INSTRUMENT LANDING SYSTEM

The ILS is the current standard system of approach throughout the world. It consists of two ground installations that transmit narrow signal beams near the touchdown point on a runway. The glide-slope beam provides the proper angle of approach and a localizer beam provides direction. The localizer beam intersects the glide-slope beam vertically to define a glide path that leads the pilot straight to the centerline of the runway at an angle of between two and six degrees. His distance from the runway is indicated by two beacons called fan markers, one at about five miles and the other at one mile from touchdown point.

MICROWAVE LANDING SYSTEM

This latest advance in airport technology may eventually replace the ILS system. Two MLS ground stations, analogous to those of the ILS, transmit scanning beams that fan out in an arc 15 miles wide and to a height of 20,000 feet. This expanded flight channel allows access to several approach paths so pilots can avoid noise-sensitive areas—as well as other aircraft.

Waiting to land
To cope with congestion around busy airports, the approach controllers may instruct incoming aircraft to circle over a holding point, usually a radio beacon about 50 nautical miles from the control tower. Each new arrival is assigned a progressively higher altitude, thereby forming a stack which can reach to 23,000 feet. The airplanes descend in steps of 1,000 feet, waiting their turn to leave the bottom of the stack at 6,000 feet. They then descend to 3,000 feet for a final approach to the runway. Commercial airliners have reserve fuel to allow for this holding period, which can last up to 30 minutes at busy airports.

Glide-slope beam

ILS VHF localizer transmitter
Provides lateral guidance to the pilot. Analogous to the ILS localizer, but offering a much wider proportional guidance coverage, is the MLS Azimuth Station, also located at the end of the runway.

ILS inner marker
Usually located about a half mile from the runway; by now the plane should be about 200 feet off the ground and 20 seconds from touchdown.

ILS outer marker
At this point, about 5 miles from the runway, the plane must be stabilized on the glide-slope and localizer beams.

Localizer beam

ILS UHF glide-slope transmitter
Beams a signal that provides vertical guidance to the pilot during his approach. Situated just in front of the ILS glide-slope transmitter and providing a wider range of glide path angles is the MLS Elevation Station.

Glide path

MLS curved initial approach
This is one of several initial approach patterns available to the pilot using the MLS.

Using Distance Measuring Equipment (DME), a pilot can read his distance to the runway from any point on the MLS signal area.

The navigational beams of the ILS are received and displayed as vertical and horizontal lines on a cockpit instrument. The pilot positions his plane by keeping the lines crossed at the center of the indicator to make sure that his approach is at the right height and angle for landing.

An Airport at Sea

Cruising at more than 80 knots, the 80,000-ton U.S.S. *Kitty Hawk* patrols the Mediterranean.

An aircraft carrier—a giant floating airbase—is one of the most formidable integrated weapon systems ever devised. It has been said that the act of maneuvering an aircraft carrier in the open seas is like steering the island of Manhattan from atop the Empire State Building.

The angled flight deck of a typical carrier is more than 1,000 feet long, 250 feet wide, and covers four-and-a-half acres. Planes are launched and retrieved every 45 seconds and in just under 500 feet, a tenth of the distance required by land-based planes. The activity on the flight deck can be so intense that it has been compared to O'Hare and Washington airports at peak hours squeezed into the size of four basketball courts.

The carrier is base to about 85 planes, each with a specific offensive or defensive role. There are all-weather fighters like the F-14 Tomcat, surveillance planes with all-seeing radar, tankers for midair refueling, both short- and long-range attack bombers, prowlers equipped with electronic jamming equipment to confuse the enemy, helicopters designed to detect and track enemy submarines or conduct search-and-rescue missions, and transport aircraft big enough to deliver entire jet engines.

For quick identification and maximum safety on the deck, the 2,000 men assigned to the carrier wear color-coded jerseys. The "air boss" and his assistant, the "mini-boss," wear yellow shirts and preside over the entire operation. Their commanding view is 140 feet above the sea in a glass-and-steel bubble called the "pri-fly" or primary flight control.

Because of the deafening noise level generated by incoming and outgoing planes, all flight-deck crew wear ear protection, and communications are coordinated by the use of hand signals. During night operations, the same signals are made with lighted wands. To further ensure the safety and efficiency of flight operations, all aircraft movement is replicated in the flight-deck control room with a scale model of the flight deck and miniature planes.

Like a shot out of a cannon, a 35-ton F-14 Tomcat blasts off a carrier's deck. In order to reach takeoff speed in such a restricted space, planes are hooked to a steam catapult system that generates over 70 million foot-pounds of energy. Even if the pilot locked his brakes, the sheer force of the thrust would hurl the plane over the bow at 130 to 150 miles per hour.

An F-14 drops like a brick out of the sky and slams onto the flight deck, its massive trailhook snagging the three-wire arresting cable, bringing the jet to a jolting halt just 30 feet from the edge of the ship. At the instant of touchdown, the pilot shoves the throttle full forward in case he misses the arresting gear. But once the tailhook snags the cable, the plane is halted in about two seconds.

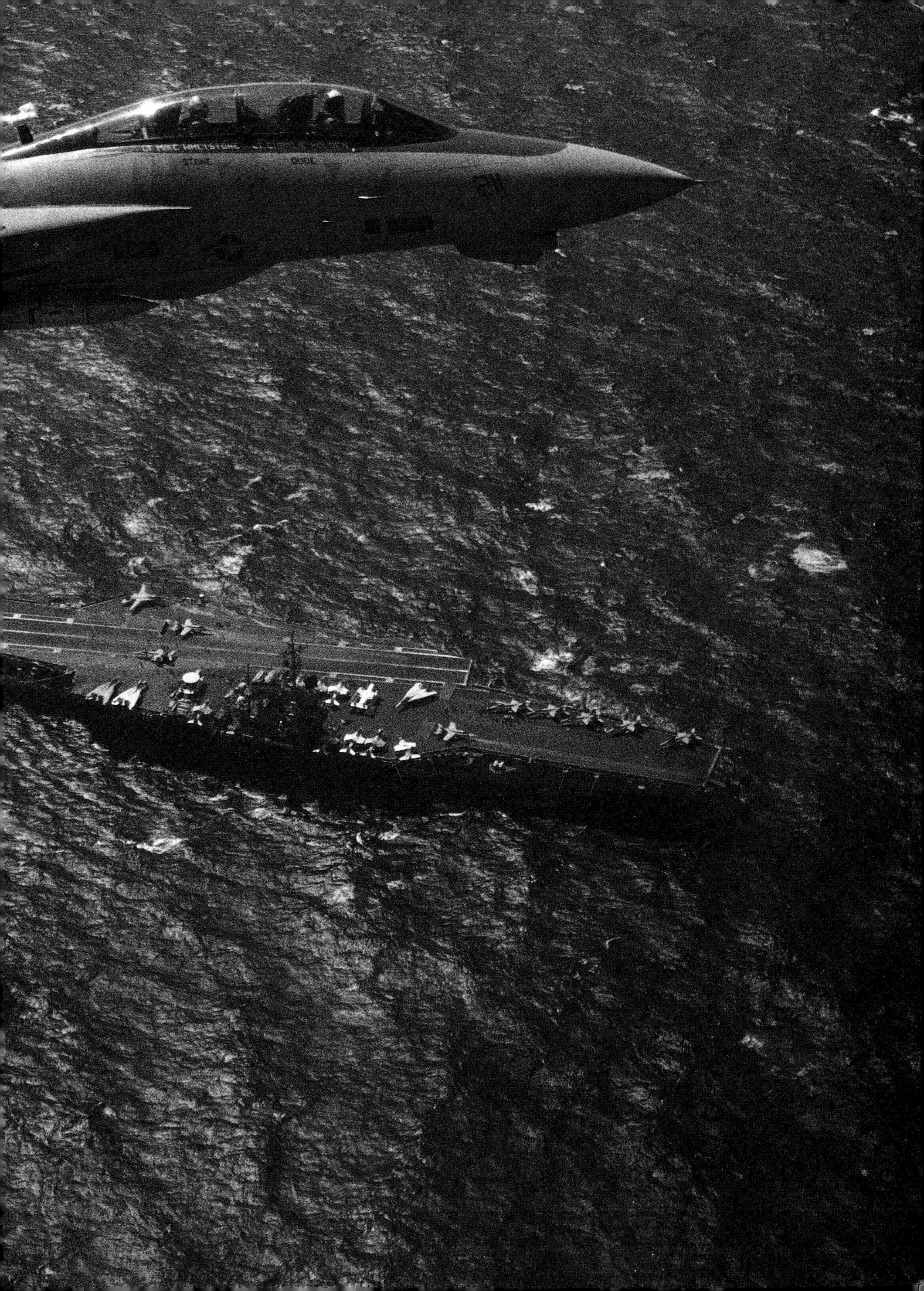
LT MIKE WHETSTONE
STONE
DUDE

QUEST FOR SPEED

It is a rare and spectacular sight, visible only in humid weather and usually only to other high-flying pilots: An F-14 jet breaches the sound barrier like an explosion bursting from one universe into another. The barrier—a normally invisible wall marking the speed at which the compressed air piling up ahead of the plane creates a sudden, massive increase in drag—reveals itself for the wink of an eye and seems to stretch with the impact of the jet's nose cone. A vaporous radiance appears; from the front of the plane it looks like a halo but from the side it presents the illusion of a battery of sharp spikes darting off the wings and fuselage. The apparent barrier tears open, repairs itself in an instant, and is gone.

Ever since the first powered flight in 1903, designers have tinkered with airfoils, engines, fuel mixtures, materials, even the number of wings to coax additional speed from their craft. Toward the end of World War II, American pilots were pushing their P-38 Lightnings to 600 miles per hour in steep, powered dives. But at those speeds buffeting became so violent that some machines literally disintegrated: Wings and tails broke away and several pilots lost their lives. The problem lay not so much in attaining the speed of sound—about 760 miles per hour at sea level—but in overcoming the severe punishment the aircraft received as it approached that speed.

In 1944, even before the end of the war, the U.S. Air Force commissioned Bell Aircraft to design a rocket-powered experimental aircraft, called the Bell X-1 —Experimental Sonic-1—that would withstand the intense pounding. At that time the only way to test such an experimental plane was to hazard flying it. A 24-year-old fighter ace, Captain Charles E. "Chuck" Yeager, was chosen as the test pilot. On October 14, 1947, the slim, bullet-shaped X-1, painted bright orange for high visibility, was launched at 20,000 feet from the bomber belly of its B-29 mother ship and headed toward an altitude of 43,000 feet. Yeager was warned not to "push the envelope" beyond 96 percent of the speed of sound unless absolutely certain that he could safely do so. As the X-1, powered by a rocket engine capable of

In this rare photograph, snapped by the pilot of a second fighter, a shock wave appears as a circle of vapor around an F-14 Tomcat as it breaches the sound barrier.

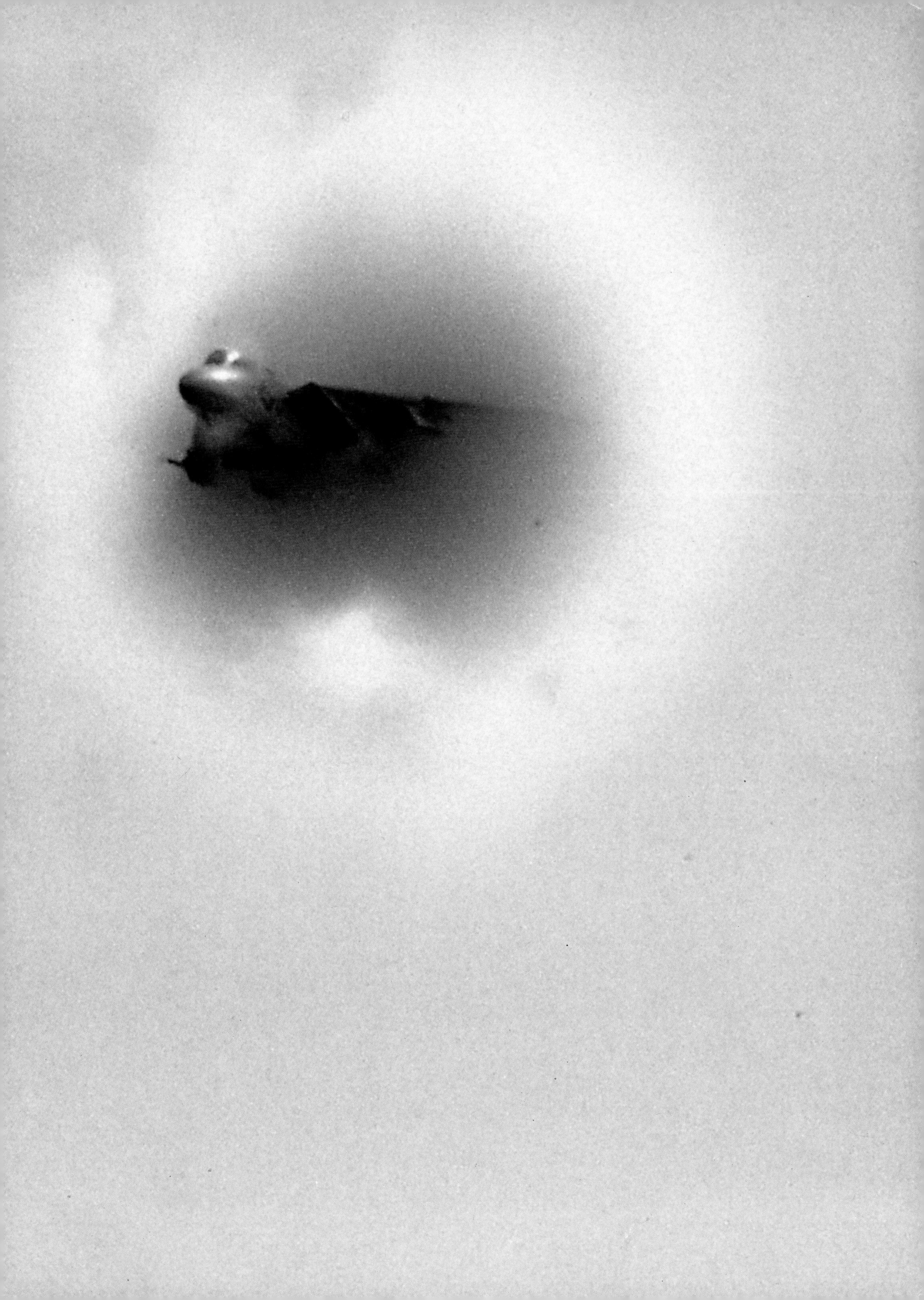

producing 6,000 pounds of thrust, climbed and accelerated, Yeager kept his eyes glued to a new device called a Machmeter, which recorded the aircraft's approach to the speed of sound.

When the meter rose above Mach .90, the machine began to shake alarmingly, nearly causing Yeager to lose control. Then, quite unexpectedly, the buffeting stopped. The Mach needle rose to Mach .965—then tipped off the scale for 20 seconds. On the ground, a tracking van reported a sound like the rumble of distant thunder: the first sonic boom made by an airplane. The data later showed that the X-1 had reached Mach 1.07—an event that Yeager later described disappointedly as a "poke through Jell-O."

BUILT FOR SPEED

By trial and error as much as by studied research, aeronautical engineers mapped the far side of the sound barrier and learned the properties of supersonic flight.

As an object—an aircraft, a bird, even a glider—moves forward through the air at any speed, it disturbs the air molecules through which it passes. They form alternating high- and low-pressure waves that radiate from the object like ripples of water around a moving boat. The pressure waves set up a vibration that to human ears becomes sound, but which in many cases is far too weak to be heard. These pressure waves always travel at the speed of sound regardless of the speed of the object.

However, the speed of sound itself is not constant but varies according to the temperature of the air. The warmer the air, the faster sound travels. At sea level in moderate temperatures, the speed of sound is about 760 miles per hour; at an altitude of 40,000 feet, where the temperature averages -60° F., its velocity drops to 660 miles per hour. To avoid ambiguity, scientists use a Mach number, named

SPEED RANGES

The shock waves generated by an aircraft traveling near, at, or beyond the speed of sound have radically changed the shape of aviation. To control turbulence at high speeds, designers sweep the wings back, so that both wings and the sonic shock waves they produce fit inside the shock cone streaming from the nose of the plane.

Subsonic: 0-Mach .80
Boeing 747: Cruises comfortably at about 620 miles per hour or Mach .82.

Transonic: Mach .80-1.2
Bell X-1: Charles Yeager was the first person to break the sound barrier, attaining a speed of Mach 1.07 in 1947.

Supersonic: Mach 1.2-5
Concorde: Cruises at 1,350 miles per hour or Mach 2.

Hypersonic: Mach 5 and above
National Aerospace Plane: Proposed transatmospheric vehicle that will take off from a conventional runway and accelerate into orbit, reaching speeds of Mach 6 and up.

for the noted Austrian physicist, Ernst Mach, who helped pioneer the study of sound. A plane's Mach number is equivalent to its speed divided by the speed of sound at the plane's altitude. Thus a plane traveling at the speed of sound—whether 760 miles per hour at sea level or 660 miles per hour at a high altitude—is said to be traveling at Mach 1, regardless of the temperature of the surrounding air.

The speed of an airplane dictates the reaction of the pressure waves to its passage. There are four main speed ranges, as shown on the chart at left: subsonic, transonic, supersonic and hypersonic. When a plane is flying at subsonic speeds—below Mach .80—the pressure waves have time to escape from the aircraft since they are traveling away from it at the speed of sound. In effect, the air particles ahead of the plane receive advance warning of its arrival from these changes in pressure and start to move out of the way. As a result, the air begins to adjust before the craft arrives, so as not to disturb its path.

As the plane approaches transonic speed—Mach .80 to Mach 1.2—the pressure waves do not have time to move out of the way of the oncoming plane since it is traveling along with them; the waves compress and the air becomes far more dense. When the plane meets the compacted air, it hits with a jolt and a series of shock waves builds up perpendicular to the direction of flight. The first shock wave attaches itself to the center of the wing's upper surface as the airflow there reaches Mach 1. As the plane's speed increases, the air under the wing also reaches Mach 1 and a second shock wave forms. Concurrently, the wave above the wing increases in size and strength.

The aerodynamic effects of these shock waves interfere with the thin layer of air that hugs the wing's surface, called the boundary layer, and result in a reduction in lift. This interference also hampers the effectiveness of control surfaces, and causes an increase in drag because the airflow behind the shock waves is turbulent. The turbulence induces shock stall, a condition that can subject conventional aircraft to violent buffeting.

As the plane accelerates to supersonic speed, the shock waves above and below the wing move back and join together at the trailing edge of the wing. Ahead of the wing, pressure is still building up and, as the plane reaches Mach 1, a new shock wave attaches itself to the wing's leading edge. Beyond the speed of sound, these waves bend back over the plane in the shape of a cone, with the nose of the plane at the cone's tip. This Mach cone trails behind like the bow wave of a ship, except that it is three-dimensional.

To solve the shock stall problem at transonic speed, aircraft designers looked at various ways of reducing the formation of shock waves. Their solutions took the shape of the DC-10, the 707 and the 747—all subsonic planes designed to cruise comfortably at Mach .80. The trick was to sweep back and "file down" the wings. In flight, air passes over these V-shaped wings at an angle, which delays the onset of shock waves until a much higher flying speed is attained. Making the wings thinner, with sharper leading edges, enabled them to slice cleanly through the compacted air.

Next was the problem of drag and turbulence beyond the speed of sound. Designers soon realized that a supersonic aircraft must fit neatly inside the Mach cone it produced. Since airflow at such speeds prefers sharp edges, the straight-winged aircraft of the World War II era were redesigned with wasp-waist fuselages, needle noses and knife-edged wings. The faster the plane, the more such streamlining it required. Thus the Lockheed SR-71 Blackbird, the culmination of the great advances made in aviation since the days of the biplanes, was given a slender fuselage and stubby delta wings that blended smoothly into chines—horizontal surfaces attached to the fuselage to increase lift. In 1974 a Blackbird set a transatlantic speed record of one hour, 54 minutes and 56 seconds for the 3,470-mile route between New York and London—less than 1/17 the time it had taken Charles Lindbergh to fly to Paris 45 years earlier. The fleet has since been retired but still holds more than a dozen speed and distance records.

In this classic photo by Harold Edgerton, a strobe light flashes for a third of a millionth of a second to freeze a bullet as it explodes through an apple at 1,800 miles per hour—Mach 2.38. The bullet's aerodynamic shape inspired early designers of supersonic aircraft, since ballistics experts agreed that the projectile's shape appeared to give it good stability at velocities above Mach 1.

Heat, too, became a problem for aircraft reaching beyond the sound barrier, where the skin temperature of an airplane can exceed 500° F. and whirling turbine blades glow at 1700° F. New products had to be created or improvised: titanium skins that could withstand inferno-like heat; corrugated panels that would expand without warping; even gold-plated electrical connections that retained their conductivity at high temperatures.

SURPRISING SUPERSONICS

The roll of thunder, the sharp report of a high-caliber bullet, even the crack of a ringmaster's whip—all are sonic booms. They are generated by shock waves—the result of sudden increases in air pressure.

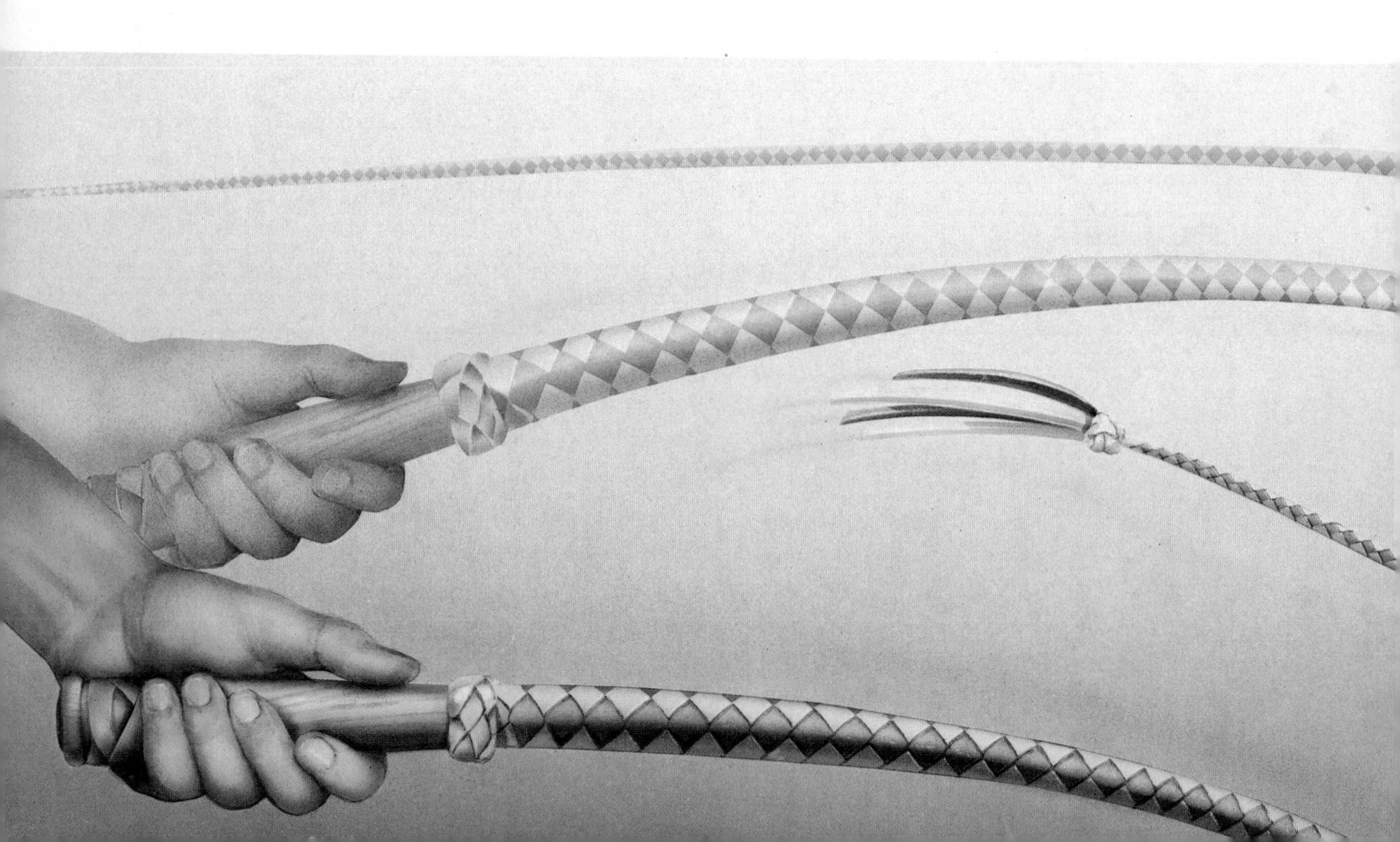

Nature's version of the sonic boom—probably the first ever heard by humans—is the thunderclap. It begins when a large, positive electrical charge builds up in the chilly upper layers of a thunderhead. Negative charges gather below and in turn set up an attraction with positive charges on the ground. At first no cloud-to-ground discharge of lightning can occur, because the air acts as an insulator. Eventually, though, a bolt zigzags to the ground at about 60 miles per second, opening up an ionized channel through the air—in effect, acting as a wire that the electricity can follow. The main bolt that follows discharges about 100 million volts of electricity, enough to light up the houses on ten city blocks, and heats the air in its path to more than 60,000° F. The heated air expands at supersonic speed, creating a cylindrical shock wave that is heard as a violent rumble. Although aeronautical engineers have succeeded in designing planes that can withstand the stress of high-speed flight, they have not eliminated its troublesome by-product: the sonic boom.

The shock wave or cone that trails behind a supersonic plane eventually dissipates above it but also touches the ground below, some time after the plane has passed. The human eardrum vibrates in response to this dramatic change in air pressure; the vibrations are carried to the brain as electrical signals and interpreted as sound. The shock wave is heard as a loud, sudden bang—the sonic boom. It lasts for only a fraction of a second, but because the change in pressure is so sudden, the sound can be startling. Anyone within the shock wave's path will experience the sonic boom, a person in another plane flying at subsonic speed, a balloonist or an observer on the ground who has seen the plane pass by seconds before the sound is heard. Concorde's signature is a distinct double sonic boom—one from the nose and one from the tail. Only the fortunate passengers for whom the sonic booms are created do not hear them, since the shock waves travel away from them.

THE CRACK OF A WHIP

As a 12-foot bullwhip snaps, kinetic energy travels from the handle along the length of the whip to its tip, which gains speed and momentum as the whip unwinds. As the energy reaches the tip threads, called the "popper," the threads are momentarily accelerated to about 1,400 feet per second, or 900 miles per hour—Mach 1.26—and produce a shock wave that in turn produces a miniature sonic boom.

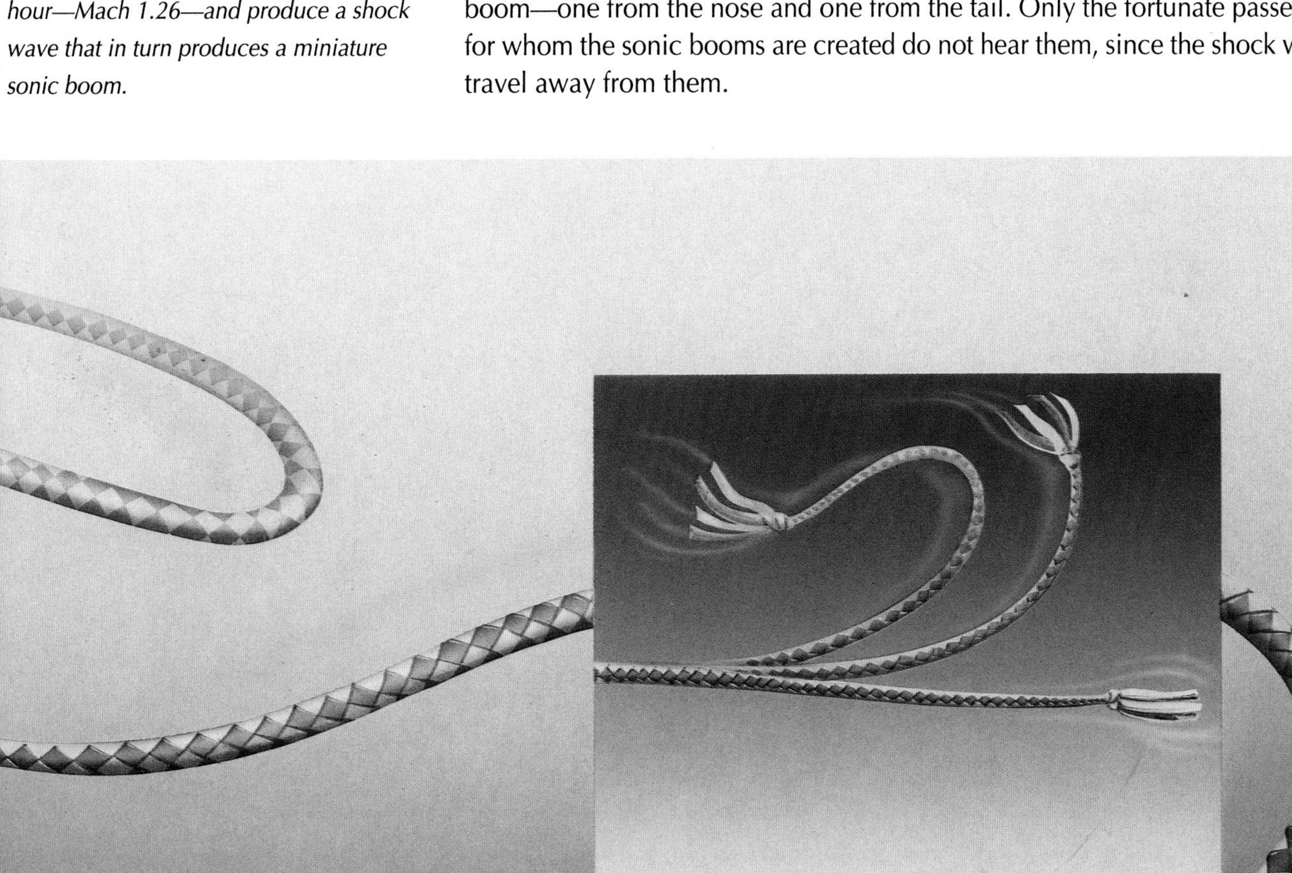

Flight Log: Concorde

Every line, curve and twist of the supersonic Concorde's 84-foot wingspan and 204-foot-long fuselage represents a precise marriage of form and function. The jet's wing, an aerodynamic compromise between the requirements of high- and low-speed flight, is swept back in a double delta compound curve known as an ogive.

Slung beneath it in two box-like structures are four Rolls-Royce turbojet engines, capable of producing 152,000 pounds of thrust. Thirty-four thousand gallons of fuel are stored in wing tanks. During flight, a system of pumps and valves shifts part of the fuel's weight to auxiliary tanks located fore and aft to compensate for changes in the plane's center of gravity.

At its cruising altitude of 50,000 feet, the Concorde's aluminum skin reaches a temperature of 262° F., and the fuselage stretches four inches, causing the windows to shift slightly. Looking forward from the flight deck, the pilot sees the horizon 300 miles away; the view is 270 degrees, spanning 250,000 square miles. Inside the Concorde's rapier-slim fuselage, up to 144 passengers are seated four abreast, where they can sip champagne and eat caviar while watching the jet's speed climb to Mach 2 on small screens called Machmeters.

0:00
TAKEOFF
Its nose tilted 5 degrees downward for better visibility, the Concorde lifts off the runway at New York's Kennedy Airport at 250 miles per hour.

0:15
SUBSONIC FLIGHT
Climbing 3,000 feet per minute, the Concorde temporarily levels off at 29,000 feet. Until the aircraft is clear of land, it will fly at 650 miles per hour—Mach .95.

0:20
ACCELERATION
Moving through the air, the Concorde produces sound waves that ripple outward like concentric spheres. At subsonic speeds, these waves move ahead of as well as behind the plane.

0:23
BREAKING THE BUBBLE OF SOUND
As the Concorde climbs to 50,000 feet and accelerates to the speed of sound (685 miles per hour at 30,000 feet), it catches up with its own pressure waves. The result: These disturbances pile up into a shock wave that is perpendicular to the line of flight.

During takeoff, the Concorde's nose must be dropped 5 degrees for the pilot to see the runway *(above)*. Once airborne, the nose is aligned aerodynamically with the fuselage in preparation for supersonic flight *(below)*. Because of its steep angle of attack, the Concorde's nose must be dropped 12.5 degrees during landing.

0:50
SUPERSONIC CRUISING
Beyond Mach 1, the pressure waves fold back over the Concorde in the shape of a cone. Where this cone touches the Earth's surface, the sudden change in pressure causes a sonic boom. At its cruising speed of Mach 2 (1,350 miles per hour), the Concorde covers one mile in less than three seconds.

2:58
SLOWING DOWN
Some 28,000 feet over Bristol, about 100 miles west of London, the Concorde slows to Mach .95 to avoid producing a loud sonic boom over southern England.

3:25
LANDING
Three-and-a-half hours and 3,500 miles from New York, the Concorde touches down at Heathrow Airport at 185 miles per hour.

For the time being, sonic booms are an environmental price for supersonic flight. Since the impact of the boom varies with altitude and aircraft design, measures can be taken to minimize the effects. A supersonic craft flying at an altitude of more than 40,000 feet creates a tolerable boom; the same plane traveling below 10,000 feet—too low to allow the shock waves to dissipate—causes an intense boom that may even shatter windows. Further streamlining of supersonic craft also helps to reduce the formation of shock waves, lessening the boom to a dull rumble.

MAN VERSUS SPEED

The final obstacle in the conquest of speed is not the brilliance of technology, but the vulnerabilities of the human body. In the whole history of aviation, only three men have "gone supersonic" without benefit of an aircraft. In separate incidents, two pilots have bailed out of planes traveling at supersonic speeds and lived to tell about it. A third set a world record for the longest delayed parachute drop and went supersonic in the process. In 1960, Captain Kittinger of the USAF stepped out of a balloon at 102,800 feet and dropped 16 miles before deploying his chute. For an instant—in rarefied air at 90,000 feet—he was plummeting earthward at a speed of 825 miles per hour, or Mach 1.25.

Wearing an anti-G suit and strapped tightly into the ejection seat of an F-15 Eagle (below), *a pilot's G-force tolerance is increased by about 1.5 Gs, raising his overall limit to about 8 Gs for 10 seconds during a high-speed maneuver.*

Unvarying speed, no matter how high, creates little stress on the human body, but sudden acceleration, deceleration or change of direction exerts a powerful, even violent force. Such a force is expressed in terms of the equivalent force exerted by the Earth's gravity at sea level (1 "G," as scientists say) and is experienced primarily as a change in body weight.

At 1 G, body weight is normal. At 2 Gs, effective body weight is doubled, and blood has difficulty reaching the head and eyes, causing cloudy vision. At 3 Gs, the legs are too heavy to lift, and peripheral vision starts to fade. By 6 Gs, arms cannot be raised above the head, and vision deteriorates from "gray-out" to blackout. A flier in excellent physical condition can "fight the G," warding off unconsciousness for the few seconds necessary to complete a maneuver by tensing the muscles of his legs and torso in an attempt to hold the blood in his head. But at 8 Gs, the blood circulation in the brain is so dangerously reduced that a pilot without a so-called anti-gravity suit will lose consciousness.

The anti-G suit *(opposite)* acts much like a corset. Bladders in the suit, located over the calves, thighs and abdomen, automatically fill with air as G-force increases, and exert a counter-pressure on the legs and abdomen, forcing some of the blood back up to the head and eyes. Reclining seats also reduce the vertical distance blood has to travel. But the suit's value is limited: It affords only about 1.5 Gs of protection at high G-forces.

An F-15 fighter pilot keeps a watchful eye on both the "enemy" and his instrument panel (above) *using a holographic heads-up display (HUD). The transparent screen, mounted at eye level, projects flight information directly to his line of sight.*

The effects of the G-forces produced by a sharp turn in a speeding jet fighter have been described this way by a pilot: "An invisible force slams my helmeted head back, punching the air from my lungs and crushing my body. My internal organs are beginning to become unglued, separating layer by layer like reluctant Velcro, and I am suddenly acutely aware of the fluids in my body. I turn all my energies to keeping myself from coming apart at the seams. I have been rendered to jelly." The tighter the turn and the higher the speed, the greater the force. Even at only 450 miles per hour, a turn with a radius of one-half mile produces 5 Gs.

Military jets such as the F-16 are crammed with computers that perform multiple high-speed tasks and permit their pilots to maneuver at supersonic speeds and paralyzing G-forces. The cockpits resemble TV control rooms, with computer-generated displays replacing mechanical dials. Fighter pilots rely on these on-board computers to constantly report on the jet's speed, angle of attack and other flight attitudes, as well as to receive guidance from external navigation sources. Backed up by multiple computers that constantly monitor one another, the system senses any deviation from desired flight conditions and responds by sending out as many as 40 electronic instructions per second to the control surfaces—far more than any human pilot could carry out. The system can even override a command that is potentially dangerous to the plane or pilot, such as a turn that would generate excessive G-force.

Another feature of supersonic jets is a computer-generated cockpit display, aptly called the heads-up display, or HUD *(left)*. Virtually everything a pilot needs to know to fly and fight—his speed, heading, his weapons systems—is continually updated by computer and projected onto a transparent glass screen mounted at eye level. The saving of those precious few seconds previously needed to glance down at the instrument panel can mean the difference between success or failure, life or death, in a bomb run or dogfight.

FLY-BY-WIRE FIGHTER

To cope with supersonic speeds and the tremendous G-forces they exert, combat fighters like the F-15 Eagle below are constructed with an exotic superalloy skin stretched over a rigid airframe, which may be carved from a solid block of metal. The nose is crammed with as many as four flight computers, and the wings are essentially huge fuel tanks. The pilot does not directly "fly" the plane in a conventional sense; instead he "commands" the flight computers to execute the kinds of maneuvers illustrated below.

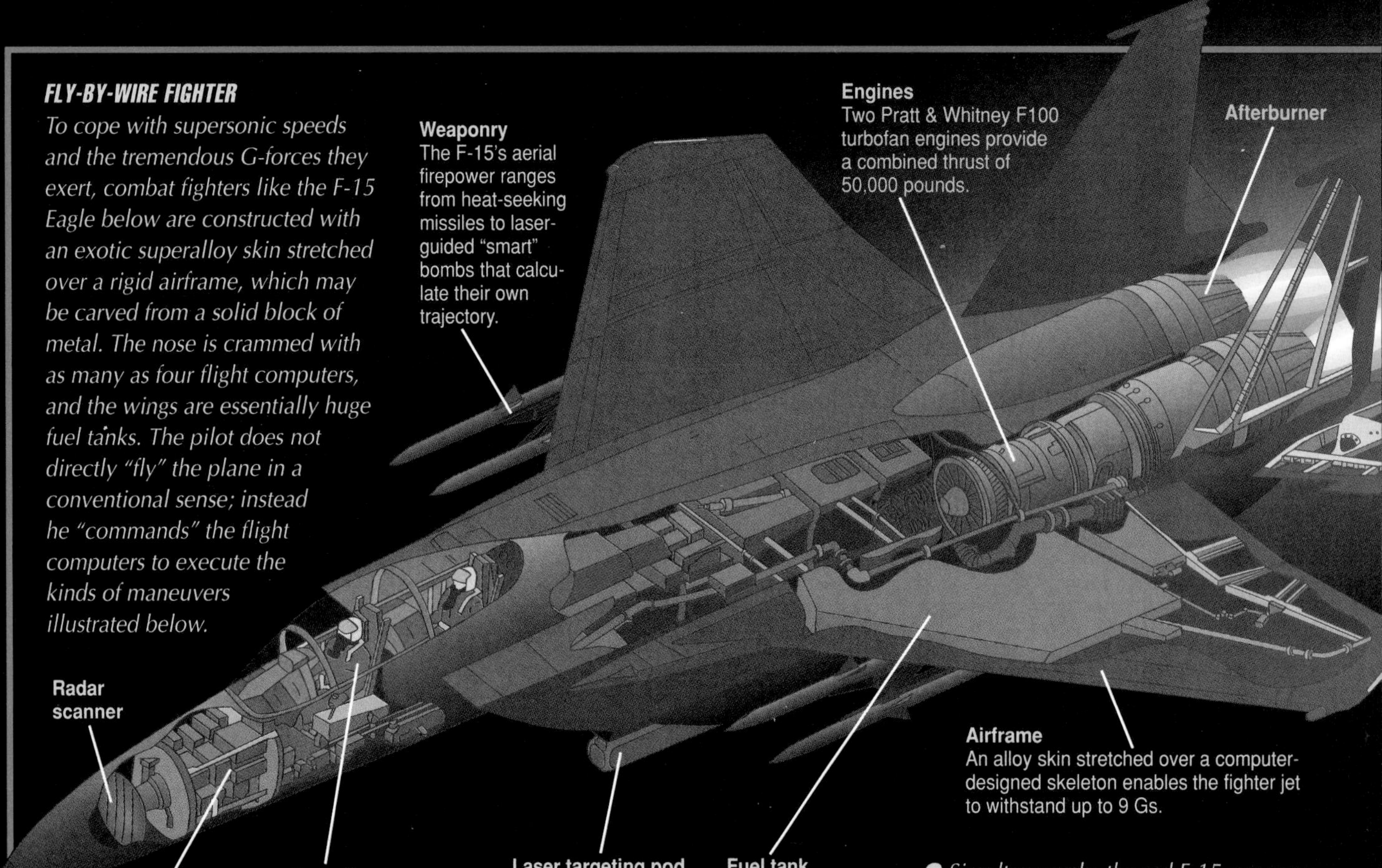

Weaponry
The F-15's aerial firepower ranges from heat-seeking missiles to laser-guided "smart" bombs that calculate their own trajectory.

Engines
Two Pratt & Whitney F100 turbofan engines provide a combined thrust of 50,000 pounds.

Afterburner

Radar scanner

Airframe
An alloy skin stretched over a computer-designed skeleton enables the fighter jet to withstand up to 9 Gs.

Electronics bay
Advanced navigation and weapons systems include terrain-following radar and infrared night vision.

Cockpit
Two crewmen are required to fly an F-15 in combat. The pilot handles the flight controls and navigation equipment while a weapon-systems officer monitors the complex radar and weaponry.

Laser targeting pod
Capable of quickly detecting long-range targets under all-weather conditions.

Fuel tank
The F-15 has a range of about 2,500 miles without refueling.

6 *Simultaneously, the red F-15 secures a good lock and "squeezes off" one short-range, heat-seeking missile. As the bogeys become tiny, visible specks to the Eagle pilots, the missile destroys one enemy aircraft.*

Dogfight at Mach 1

1 *An Airborne Warning and Control System (AWACS) aircraft 100 miles away directs three F-15 Eagles* (above) *to intercept two bogeys—unidentified hostile aircraft—20 miles distant and closing in at the speed of sound* (upper right). *Traveling in a chain, the Eagles accelerate to Mach 1 just as they are detected on enemy radar.*

The Korean Conflict of the early 1950s was the first jet air war. For the next four decades, new technologies and maneuvering techniques were put to the test in the skies above Vietnam, Israel and the Persian Gulf. Distances between fighters increased from several hundred yards to several miles in compensation for the much-faster speed and larger turning radius of the new aircraft. The Vietnam War saw the advent of the air-to-air (AAM) missile. The dramatic increase in range of the AAM over the gun rendered traditional fighting wing formations obsolete.

Shown in numbered steps on these pages is a simplified version of what a dogfight of the future might resemble. Because of the distances and speeds involved, it is as much a battle of computers as of fighter pilots.

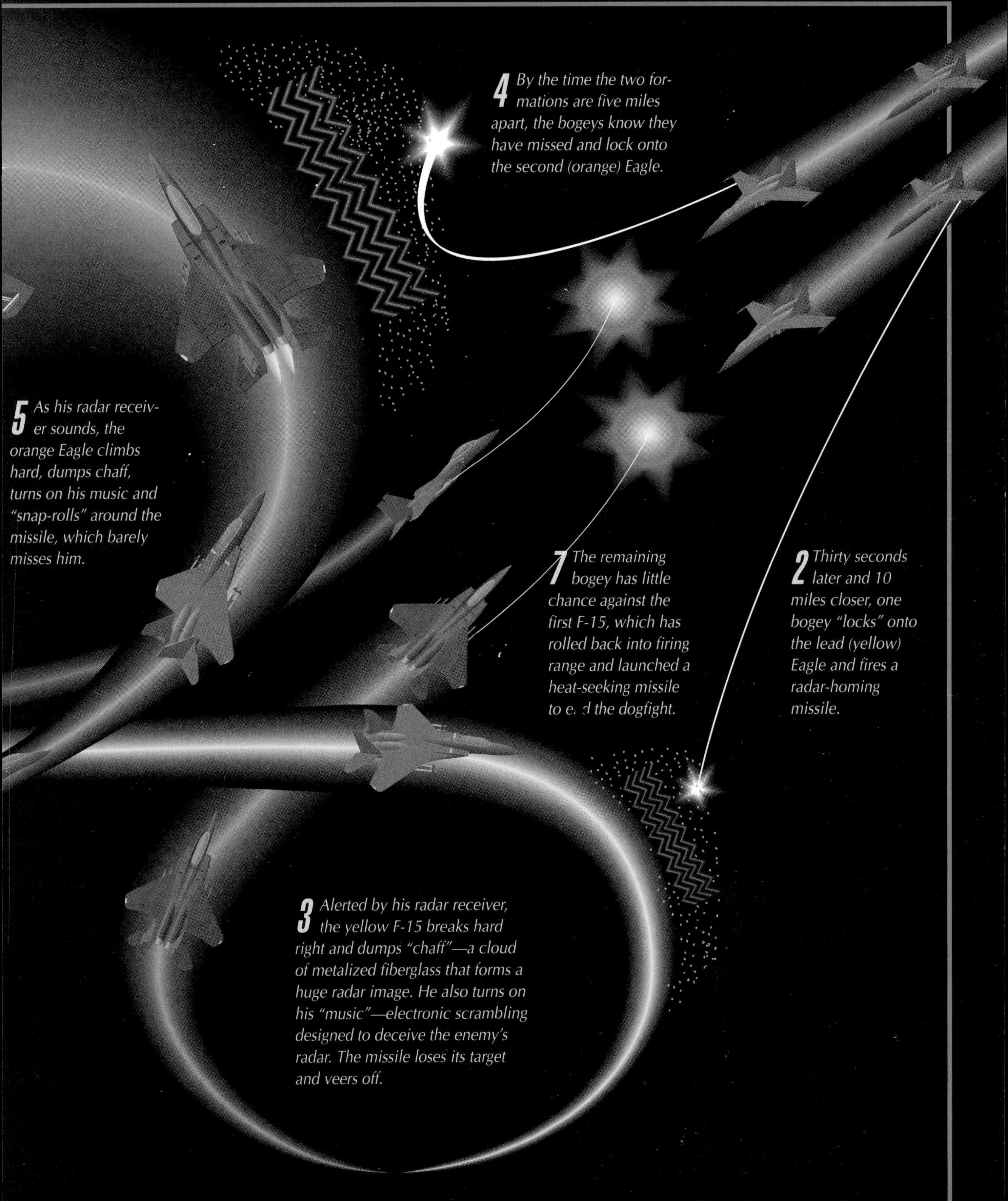

2 Thirty seconds later and 10 miles closer, one bogey "locks" onto the lead (yellow) Eagle and fires a radar-homing missile.

3 Alerted by his radar receiver, the yellow F-15 breaks hard right and dumps "chaff"—a cloud of metalized fiberglass that forms a huge radar image. He also turns on his "music"—electronic scrambling designed to deceive the enemy's radar. The missile loses its target and veers off.

4 By the time the two formations are five miles apart, the bogeys know they have missed and lock onto the second (orange) Eagle.

5 As his radar receiver sounds, the orange Eagle climbs hard, dumps chaff, turns on his music and "snap-rolls" around the missile, which barely misses him.

7 The remaining bogey has little chance against the first F-15, which has rolled back into firing range and launched a heat-seeking missile to end the dogfight.

The Leading Edge

An electronics specialist makes final adjustments to QN's delicate autopilot—the "brain" of the unusual model—in preparation for a test flight.

Bridging time, a replica of the prehistoric flying reptile Quetzalcoatlus northropi *swoops over Racetrack Dry Lake in Death Valley, California.*

In 1986, aviation aeronautical engineer Dr. Paul MacCready and his team of scientists gathered on a dry lakebed in Death Valley, California, for the debut of a curious flying machine. The fur-covered, electronics-filled contraption was a remarkably lifelike replica of *Quetzalcoatlus northropi*, the reptilian pterodactyl of 100 million years ago.

If MacCready and his team managed to fly "QN," as they called their contrivance, it would be more than just a bit of high-tech whimsy; never before had science achieved the flapping flight of birds—or pterodactyls. The replica was a self-contained, fly-by-wire model incorporating the latest advances in alloys and plastics, computers and robotics. Its electric-motor "muscles" were powered by six pounds of batteries and directed by an electronic "brain" that received instructions from a controller on the ground, an on-board gyroscope and a wind vane incorporated in its neck. The 18-foot wings, made of a carbon-fiber skeleton covered with latex, could flap at varying speeds, move backward and forward to control lift, and twist or "warp" to counter the effect of side winds.

QN was towed into the sky on a winch-powered sled. Once airborne, the platform was jettisoned and floated earthward on a tiny parachute. On the ground, the controller flicked a switch to activate QN's autopilot and 13 flight motors. Five hundred feet overhead, QN slowly wagged its wings and, as MacCready and his team held their breath and squinted skyward, soared and flapped for three glorious minutes before landing.

Otto Lilienthal tests one of his cotton-and-willow-rod gliders. Although he appears to be out of control, he has just bent his knees to adjust the glider's center of gravity.

A hang glider rides a thermal across the slopes of Mount Hood in Oregon. Unlike Lilienthal's manned kites, modern hang gliders are equipped with a harness and crossbar that allow the pilot to steer by shifting his weight.

The modern sport of hang gliding was born of science and nurtured in serendipity. One of the earliest pioneers was the great German engineer Otto Lilienthal, shown sailing from a hill near Berlin in the late 1890s. Lilienthal believed that gliding, not flapping, would be the basis for powered flight, and to prove his contention he built and maneuvered around in hundreds of wood-and-fabric gliders until a tragic accident ended his investigations. Some of the flights were very dashing, though sport was the farthest thing from Lilienthal's mind.

More than half a century later, another engineer, NASA's Dr. Francis Rogallo, carried the idea of hang gliding a large step forward—again in the cause of science. Rogallo designed a remote-controlled kite that could be deployed to bring space capsules safely back to earth once they had reentered the atmosphere. Wind-tunnel testing confirmed that the delta-shaped wing possessed both aerodynamic stability and remarkable lifting powers. In the end, NASA found other ways to bring home its astronauts and satellites. But Dr. Rogallo's elegant and inexpensive wing, meant for space travel, has now made its mark as the basic design for the burgeoning sport of hang gliding.

Since the 1970s perhaps a million people worldwide have taken up this least expensive and most accessible form of flight. A typical flight begins atop a high ridge with strong prevailing winds. The pilot wears a shoulder-to-knee harness, which is in turn hitched to the frame of the glider. Facing into the wind, the pilot runs forward until the "sail" fills with air and lifts him skyward. To control the glider, he holds onto the crossbar and shifts his weight from side to side. To slow down and land, he pushes the bar forward.

An experienced pilot, with a keen eye for rising thermals, may climb to a height of 10,000 feet and stay aloft for ten minutes or more. Otto Lilienthal, who achieved glides of about 1,000 feet, would have been amazed—and delighted.

Wars and the threat of war have given impetus to great leaps in aviation technology, sometimes at a pace too fast for the success of a design.

Jack Northrop, a high-school graduate who taught himself engineering before founding the firm that now holds his name, postulated that the most graceful and efficient shape for an aircraft was a flying wing. Among other benefits, the absence of a fuselage would reduce both drag and structural weight. Starting in World War II, a succession of promising designs led ultimately to the first flight in 1947 of the Northrop YB-49, a flying-wing bomber powered by eight turbojet engines *(inset)*. The design, although at least 20 years ahead of its time, was dogged by stability, control and mechanical problems. These—and, some sources say, political chicanery—led to the cancellation of the government contracts.

It was a bitter disappointment for Jack Northrop, but before his death in 1981, his dream came alive again in designs for the ultra-secret B-2 bomber. The B-2's unusual shape combines large interior volume—to house weapons, fuel and avionics—with a radar-avoiding profile. Ironically, the 172-foot wingspan of Northrop's new high-tech warbird is exactly the same as its untimely forebear, the flying wing.

The ill-fated Northrop YB-49 was the first tailless aircraft. In 1949, the 172-foot-wide "flying wing" set a world record by remaining airborne for more than nine hours without refueling.

The bat-winged B-2 bomber flies over Palmdale, California, the home of the Northrop plant where it was built.

MCDONN

In this artist's rendition a high-speed civil transport (HSCT) cruises over Florida at more than three times the speed of sound.

Despite its age, the Concorde remains an inspiring design. Lifting off the runway at 250 miles per hour, this aircraft will reach the speed of sound in 23 minutes.

Passengers on board the Concorde are pampered for four hours as they speed across the Atlantic at Mach 2. Only military pilots and astronauts have flown faster—but without the creature comforts of leather seats and champagne. The Concorde fleet is aging, however, and designers are developing commercial airplanes that can fly even faster, farther and higher.

Fly-by-wire technology, in which on-board computers interpret the pilot's commands to adjust speed, direction and altitude, is replacing mechanical control systems. Lightweight plastics, carbon fiber and other space-age materials are already appearing in wings and fuselages. Early in the next century, jet fighters may even be flown in combat via remote control by pilots on the ground.

Armed with new technologies, companies such as Boeing and McDonnell Douglas are actively pursuing the idea of high-speed civil transports (HSCTs) like the design at left. But flight above Mach 3 poses new challenges. Since HSCTs would cruise above 60,000 feet, their engines must emit less nitrogen oxide than conventional jets so as not to harm the ozone layer. To be commercially viable, the new aircraft would require a range of 7,000 miles, twice that of the Concorde. Designers must also find ways of limiting sonic boom in order to win permission for supersonic flight over land.

There is no question, given the rapid march of science, that these problems will be solved—and that airlines may one day in the not-too-distant future carry passengers between New York and Tokyo in just over three hours.

Index

Numerals in *italics* indicate an illustration of the subject mentioned.

Q-R

S

T

U-V

W

Y-Z

PICTURE CREDITS

Credits are read from left to right, from top to bottom by semicolons.

Front cover: Bruce Dale © National Geographic Society.

6 AllSport/Vandystadt/West Light; Tom & Pat Leeson; Alain Guillou. 7 C.J. Heatley III (2); Courtesy Rockwell International. 14, 15 Alain Guillou. 18, 19 Dave Becker/Viewfinders (5). 22 Gene Stein/West Light. 23 David Lawrence/The Stock Market. 26 Julie Ades/Visions. 30, 31 C.J. Heatley III. 32, 33 Stephen Dalton/Photo Researchers. 36 EMI Pathé News Library. 37 © S.C. Johnson & Son/Smithsonian Institution, courtesy IMAX Systems Corporation. 40, 41 Mark Greenberg/Visions (2). 44 NASA/Science Source/Photo Researchers. 45 NASA/Langley. 53 Charles O'Rear/West Light. 57 C.J. Heatley III; Barry Griffiths/Photo Researchers; Mark Greenberg/Visions; M.P. Kahl/DRK Photo; C.J. Heatley III; Roy Morsch/Bruce Coleman Inc. 58, 59 Courtesy Boeing Commercial Airplane Co. (3); Gabe Palmer/The Stock Market. 60, 61 Courtesy Boeing Commercial Airplane Co. (2); Philip C. Jackson (2). 62 Courtesy Boeing Commercial Airplane Co. 64, 65 James Sugar/Black Star. 66, 67 Stephen Dalton/Photo Researchers. 81 Tom & Pat Leeson. 82, 83 Courtesy Nikon Corporation. 84, 85 Peter Thomann/Stern/Black Star; Courtesy Superflight Inc. 92 Dr. Marvin Luttges/BioServe Space Technologies (2). 94, 95 Stephen Krasemann/DRK Photo. 96, 97 Ron Austing/Photo Researchers; Jeff Foott. 104 Chris Sorenson/The Stock Market; Ron Watts/First Light. 105 Chris Sorenson/The Stock Market. 107 Bob Klein, courtesy Unisys Corporation. 110 Courtesy Honeywell Inc., Sperry Flight Systems Group. 113 Courtesy Honeywell Inc., Sperry Flights Systems Group. 117 Allen Green/Science Source/Photo Researchers. 118, 119 C.J. Heatley III (3). 120, 121 C.J. Heatley III. 124 Dr. Harold J. Edgerton/Palm Press Inc. 128 Courtesy Boeing Military Airplanes. 129 George Hall. 132, 133 © S.C. Johnson & Son/Smithsonian Institution, courtesy National Air & Space Museum; © S.C. Johnson & Son/Smithsonian Institution, courtesy IMAX Systems Corporation. 134, 135 Bill Ross/West Light; National Air & Space Museum/Smithsonian Institution. 136, 137 Courtesy Northrop Corporation (2). 138, 139 Courtesy McDonnell Douglas Corporation; Yan Lukas/Photo Researchers.

ILLUSTRATION CREDITS

Credits are read from left to right, from top to bottom by semicolons.

8, 13 Robert Monté. 16 Steve Louis. 20, 21 Ronald Durepos. 27 Luc Normandin. 28, 29 Luc Normandin. 34, 35 Josée Morin. 52 Gilles Beauchemin. 62, 63 Gilles Beauchemin. 80 Gilles Beauchemin. 87 Josée Morin. 90, 91 Luc Beauchemin. 92, 93 Gilles Beauchemin. 98, 99 Luc Normandin. 102, 103 Guy Charette. 106, 113 Guy Charette. 114 Jean-Claude Gagnon. 116, 117 Guy Charette. 122, 123 Robert Monté. 124 Steve Louis. 126, 127 Pyrate Communications Inc.

ACKNOWLEDGMENTS

The editors wish to thank the following:
Alan Adler, Superflight, Inc., Palo Alto, CA; Delise Alison, Redpath Museum, McGill University, Montreal, Que.; Mark Allen, Mark Allen Productions, Oyster Bay, NY; Dr. Allison Andors, American Museum of Natural History, New York, NY; Ted Bailey, General Electric Aircraft Engines, Cincinnati, OH; Barkley E. Bates, Voyager Airways, North Bay, Ont.; Dr. Craig Bohren, Department of Meteorology, The Pennsylvania State University, University Park, PA; Christiane Brisson, Public Affairs, Air Canada, Dorval, Que.; Dr. Jim Britten, Department of Chemistry, McGill University, Montreal, Que.; British Airways, Toronto, Ont.; John Burton, Experimental Aircraft Association, Oshkosh, WI; Steve Buss, Experimental Aircraft Association, Oshkosh, WI; Gary Butcher, Porsche Aviation Products Incorporated, Reno, NV; Thomas R. Cole, Boeing Commercial Airplane Group, Seattle, WA; Martyn Cowley, Aerovironment Inc., Monrovia, CA; Mark Drela, Massachusetts Institute of Technology, Cambridge, MA; Abdul Elkeesh, St. Hubert Airport, St. Hubert, Que.; Carolyn M. Fennell, Orlando International Airport, Orlando, FL; Festival de Mongolfières du Haut-Richelieu, Quebec; Alan E. George, ILC Dover, Inc., Frederica, DE; John Gerike, Boeing Defense Space Group, Seattle, WA; Danielle Gerrard, Boeing Commercial Airplane Group, Seattle, WA; Dr. David L. Gibo, Department of Zoology, University of Toronto, Toronto, Ont.; Edward C. Gorman, Honeywell Air Transport Systems Division, Phoenix, AZ; Serge Guilbault, Montreal Area Control Center, Dorval, Que.; Richard Harrison, Boomerang Man, Monroe, LA; H. Keith Henry, NASA Langley Research Center, Hampton, VA; Dr. Frank H. Heppner, Department of Zoology, University of Rhode Island, Kingston, RI; Alain Jacques, Dorval Control Tower, Dorval, Que.; Linda Justo, Honeywell Inc., Business & Commuter Aviation Systems Division, Glendale, AZ; Charles Larocque, Bell Helicopter Textron Canada, St. Janvier, Que.; Michel LeBrun, Ecole Nationale d'Aérotechnique, St. Hubert, Que.; Pierre Legault, Bell Helicopter Textron Canada, St. Janvier, Que.; Captain K.D. Leney, British Airways (Concorde Division), Heathrow Airport, London, U.K.; Richard Leyes, Aeronautics Department, National Air and Space Museum, Smithsonian Institution, Washington, DC; Paul T. MacAlester, Hillsborough County Aviation Authority, Tampa, FL; Dr. Paul MacCready, Aerovironment Inc., Monrovia, CA; Vernon C. Maine, Kollsman, Merrimack, NH; J. Campbell Martin, NASA Langley Research Center, Hampton, VA; Judy Mills, Institute for Aerospace Studies, Downsview, Ont.; Bruce Nesbitt, Bell Helicopter Textron Canada, St. Janvier, Que.; Carol Petrachenko, NASA Langley Research Center, Hampton, VA; Pratt & Whitney Canada, Longueuil, Que.; Elizabeth V. Reese, Boeing Commercial Airplane Group, Seattle, WA; Richard F. Saler, The Goodyear Tire & Rubber Company, Akron, OH; Richard Shoemaker, Montreal, Que.; The Intrepid Sea-Air-Space Museum, New York, NY; Anna C. Urband, Navy Office of Information, Washington, DC; Captain Daniel P. Whalen, USN, Department of the Navy (Air Warfare), Washington, DC.

The following persons also assisted in the preparation of this book:
Dominique Gagné, Shirley Grynspan, Stanley D. Harrison,
Carolyn Jackson, Brian Parsons, Shirley Sylvain.

This book was designed on Apple Macintosh® computers, using QuarkXPress®
in conjunction with CopyFlow® and a Linotronic® 300R for page layout and
composition; Stratavision®, Adobe Illustrator 88® and Adobe Photoshop® were
used as illustration programs.

Flight. -- Time-Life Books, c1990.
(How things work)

1. Flight - Popular works. I. Time-Life Books.
II. Series.